The Impact of Tourism in East Africa

TOURISM AND CULTURAL CHANGE

Series Editors: Professor Mike Robinson, *Ironbridge International Institute for Cultural Heritage, University of Birmingham, UK* and Professor Alison Phipps, *University of Glasgow, Scotland, UK*

Understanding tourism's relationships with culture(s) and vice versa is of ever-increasing significance in a globalising world. TCC is a series of books that critically examine the complex and ever-changing relationship between tourism and culture(s). The series focuses on the ways that places, peoples, pasts and ways of life are increasingly shaped/transformed/created/packaged for touristic purposes. The series examines the ways tourism utilises/makes and re-makes cultural capital in its various guises (visual and performing arts, crafts, festivals, built heritage, cuisine, etc.) and the multifarious political, economic, social and ethical issues that are raised as a consequence. Theoretical explorations, research-informed analyses and detailed historical reviews from a variety of disciplinary perspectives are invited to consider such relationships.

All books in this series are externally peer-reviewed.

Full details of all the books in this series and of all our other publications can be found on http://www.channelviewpublications.com, or by writing to Channel View Publications, St Nicholas House, 31–34 High Street, Bristol BS1 2AW, UK.

TOURISM AND CULTURAL CHANGE: 58

The Impact of Tourism in East Africa

A Ruinous System

Anne Storch and Angelika Mietzner

CHANNEL VIEW PUBLICATIONS
Bristol • Blue Ridge Summit

DOI https://doi.org/10.21832/STORCH8373

Library of Congress Cataloging in Publication Data
A catalog record for this book is available from the Library of Congress.
Names: Storch, Anne, author. | Mietzner, Angelika, author.
Title: The Impact of Tourism in East Africa: A Ruinous System/Anne Storch and Angelika Mietzner.
Description: Blue Ridge Summit: Channel View Publications, 2021. | Series: Tourism and Cultural Change: 58 | Includes bibliographical references and index. | Summary: "This book explores the relationship between imperial formations and individual encounters at African tourist sites. It examines how encounters between tourists and hosts tend to be constructed along colonial thought lines and shows that ruination is omnipresent in postcolonial tourist settings. This book is open access under a CC BY ND licence"— Provided by publisher.
Identifiers: LCCN 2020056466 (print) | LCCN 2020056467 (ebook) | ISBN 9781845418366 (paperback) | ISBN 9781845418373 (hardback) | ISBN 9781845418380 (pdf) | ISBN 9781845418397 (epub) | ISBN 9781845418403 (kindle edition)
Subjects: LCSH: Tourism—Social aspects—Africa, East. | Language and culture—Africa, East. | Sociolinguistics—Africa, East.
Classification: LCC G156.5.S63 S76 2021 (print) | LCC G156.5.S63 (ebook) | DDC 306.48190967—dc23 LC record available at https://lccn.loc.gov/2020056466
LC ebook record available at https://lccn.loc.gov/2020056467

British Library Cataloguing in Publication Data
A catalogue entry for this book is available from the British Library.

ISBN-13: 978-1-84541-837-3 (hbk)
ISBN-13: 978-1-84541-836-6 (pbk)
ISBN-13: 978-1-84541-838-0 (pdf)
ISBN-13: 978-1-84541-839-7 (epub)

Open Access

 Except where otherwise noted, this work is licensed under the Creative Commons Attribution-NoDerivatives 4.0 International License. To view a copy of this license, visit http://creativecommons.org/licenses/by-nd/4.0/ or send a letter to Creative Commons, PO Box 1866, Mountain View, CA 94042, USA.

Channel View Publications
UK: St Nicholas House, 31–34 High Street, Bristol BS1 2AW, UK.
USA: NBN, Blue Ridge Summit, PA, USA.

Website: www.channelviewpublications.com
Twitter: Channel_View
Facebook: https://www.facebook.com/channelviewpublications
Blog: www.channelviewpublications.wordpress.com

Copyright © 2021 Anne Storch and Angelika Mietzner.

All rights reserved. No part of this work may be reproduced in any form or by any means without permission in writing from the publisher.

The policy of Multilingual Matters/Channel View Publications is to use papers that are natural, renewable and recyclable products, made from wood grown in sustainable forests. In the manufacturing process of our books, and to further support our policy, preference is given to printers that have FSC and PEFC Chain of Custody certification. The FSC and/or PEFC logos will appear on those books where full certification has been granted to the printer concerned.

Typeset by Nova Techset Private Limited, Bengaluru and Chennai, India.

Contents

	Figures and Boxes	vii
	Abbreviations	ix
	Check-in	xi
1	Boarding	2
	Language and Tourism	2
	Sound and Light Show	5
	Sounds in a Village	7
	Work Trips	9
	Structure of this Book	11
2	Terrible Magical Ways of Healing	16
	Angelika Took a Plane to Mombasa	16
	Salty Water Ends Somewhere	17
	Coconut Palm Trees Make a Pleasant Sight	21
	The Truth Will Out	21
	Language Itself Is a Curse	25
	Coconut Oil	27
3	The Philosophy of *Hakuna Matata*	30
	It Is Not Easy to Leave the Beach	30
	The Genesis of *Hakuna Matata* as a Philosophy	31
	Hakuna Matata Permeates Everything	32
	Into the *Hakuna Matata* Void	36
	Moving Out of the *Hakuna Matata* World	40
	Worms Moving In	42
4	Karen	46
	A House	46
	A Woman	49
5	Highway to Hell	56
	Transition Grammar	56
	Roadside Narratives	60
6	Ruins on the Beach	68
	Old Sites	68
	Old Roads	74

7	Relocation and Relationships	82
	Encounters on the Way	82
	Sensual Stranger	83
	A Moved Village	84
	Speaking, Smelling and Settling	85
	In the Showcase	88
8	On Various Boundaries	94
	Tourism and Postcolonialism	94
	Worldwide Known Insider Facts	95
	Linguistic Landscaping	100
9	Hostility on a T-Shirt	104
	Items with Language on Them	104
	Three T-Shirts	105
	Language Course	110
10	Movies on Sex Tourism that You Shouldn't Miss	122
	Competition, Image, Desire	122
	Films	126
	Watching *Sand Dollars*	128
	Watching *Patong Girl*	130
	Watching *The White Masai*	131
	Watching *Paradise Love*	133
11	The Ancient Speaker	138
	Language and Heritage	138
	In the Exhibition: Language as Dead Matter	140
	The Ancestors: Inverted Others, Tidy Pasts	144
	Outside the Exhibition: Language of the Unlost	145
12	Cooking Class	148
	Soup and Halva	148
	Pepper Journey	152
	Poisonous Stuff	156
13	Glossy Glossary	160
	References	165
	Index	173

Figures and Boxes

Figures

Figure 1.1	Linguistic landscape referring to a pilgrim	6
Figure 3.1	Sun protective lotion with the promise of no problem	34
Figure 3.2	Swahili as the language of Africa: A wall in Christiania (Copenhagen)	38
Figure 3.3	Mombasa airport waving goodbye with best greetings	39
Figure 4.1	A vague tractor	48
Figure 5.1	Owners of a school destroying their own road sign	61
Figure 5.2	Remnants of a house partly built on governmental land	61
Figure 5.3	Remnants of another house partly built on governmental land	62
Figure 5.4	Demolished by bulldozers: the wall of a school in Tiwi	62
Figure 5.5	No houses left	63
Figure 5.6	Demolition of homes	63
Figure 5.7	What is left: coloured walls of former happy homes	64
Figure 6.1	A vague giraffe	69
Figure 6.2	Giraffe and tree	69
Figure 6.3	Elephant, trees and palms	70
Figure 6.4	Elephant and pool (dry)	70
Figure 6.5	Buffalo and rhino amidst scenery	71
Figure 6.6	Numbered elephant	71
Figure 6.7	Ad and palm tree	78
Figure 6.8	Road and pine trees	78
Figure 6.9	More pine trees	79
Figure 6.10	Pine trees and cactus	79
Figure 6.11	Windmill ruins	80
Figure 6.12	Highway bridge	80
Figure 7.1	An emblematic rake, a tool for clearing the ground	90
Figure 8.1	Indian Ocean, wall and sunbeds	96
Figure 8.2	Signs warning about Beach Boys	98

Figure 8.3 Make sure to stay at a hotel that offers beach walks once a day. Armed officers will accompany you and other hotel inmates during a two-hour lasting tour. Beach Boys have to stay in a safety zone, being only verbally able to accompany the group, what they often do by throwing in words like 'kindergarden' or the like. It is unproblematic to ignore them 99

Figure 8.4 Staying in a place of wet feet guarantees distance from the Beach Boys. The picture is a proof of the possible existence of such a place – not first encounters, but first avoidance. The Beach Boys who have the chance to install themselves between the shore and the entrance of the hotel in order to offer necklaces and dhow tours will catch you if you don't use the trick of running like hell 99

Figure 9.1 First T-shirt: Cairo 106
Figure 9.2 Second T-shirt: Zanzibar 107
Figure 9.3 Third T-shirt: Mallorca 109
Figure 10.1 Sexual geography 128
Figure 11.1 Signboard in Simon's Town, South Africa 142
Figure 12.1 Chronotopes on a menu 154
Figure 12.2 Time eating away a signboard 157
Figure 12.3 a, b & c Cutting prior to cooking 158

Boxes

Insider Tip 8.1 Security advice 97
Insider Tip 8.2 Tourist stereotypes 101

Abbreviations

1,2,5,…	Swahili noun classes
1SG	first-person singular
COMP	comparative
DEF	definite
DIR	directional
IMP	imperative
INTENS	intensifier
INTERR	interrogative
LOC	locative
MIR.VIS	visual mirative
NEG	negative
O	object
PASS	passive
PAST	past
PL	plural
POSS	possessive
PRES	present
REL	relative
SG	singular
UNESCO	United Nations Educational, Scientific and Cultural Organization
V.O.C.	*Vereenigde Oostindische Compagnie*

Check-in

Checking out, in fact. Checking out of disciplinary duress, debilitating constraints and restrictions – of how we should do and present research, represent academia and so on. Checking in to the real life that we always knew existed as well – outside imagined linguistic monotony and predictable regularity.

We happened to begin our work on tourism and language, reckless consumption and the words and silences that go along with it at around the time that a President of the United States was surprisingly elected. Just around that time, a group of prominent scholars, we recall, had published an essay on how they wondered whether they had lost connection with ordinary people, having thus been unable to foresee what now had happened. Since then, we have been even more interested in looking at precisely what we thought academia had sheltered us from some time earlier: the trivial, ruinous system of everyday practices that destroyed almost everything that should have been safe.

Academic travel became fieldwork. Our experiences and bodies turned into data, like those of other people had always done before.

We remain grateful to all our relatives, colleagues and friends who continued their conversations with us, and from whose hospitality we greatly benefited, in many ways: Alison Phipps, Antony Kapulanga, Bella Hoogeveen, Bonciana Lisanza Nasiali, Cassandra Gerber, Chris Bongartz, Christine Rottland, Janine Traber, Janna Perbix, Katrin Storch, Kiko, Louis Mietzner, Mary Chambers, Mzee Mohamed, Mohamed, Nico Nassenstein, Rogers Maasai, Sasha Aikhenvald, Sara Zavaree, Sophie Storch, Stephen, Tawona Sithole, Usi Juma. And we owe lots of thanks to Sarah Williams, Florence McClelland, Rebana Barvin and Barbara Faux for taking such good care of our book.

All photographs in this book have been taken by us.

1 Boarding

Language and Tourism

This sociolinguistic study is dedicated to the description and analysis of language and its contexts in postcolonial tourism. It is intended as a 'mosaic' of different texts on pressing issues of alienated and homeless language associated with notions of social inequality, gaps and cracks in intercultural communication, violent environments and embodied Otherness. The sites at which these descriptions and analyses are located include beaches, ruins, bodies, objects, villages, roads and TV sets. The geopolitical focus is on East Africa.

The motivation of this book has been twofold: first, there have been relatively few contributions on language in African tourism settings so far in tourism studies and, second, there has been relatively little interest in the relevance of the tourism economy for language in African linguistics up to this date. The present book is an attempt to open up new perspectives with regard to these desiderata. In the following, we provide an overview of the pre-existing work in the sociolinguistic field that has guided the development of our research and the preparation of this book.

Tourism, as one of the globe's fasted growing industries, and its extractive presence in the Global South and many of the colonized parts of the world, has attracted considerable attention and interest from linguists and others over the past few years. This interest is very often focused on English and the hegemonic colonial languages that dominate interactions in globalized tourism. As an alternative to the (very normative and Eurocentric) approach to the global use of colonial languages as 'varieties', 'world Englishes/Frenchs/Germans (etc.)', and so on, Alastair Pennycook (2001) considered the global spread of English within a framework of postcolonial performativity and proposed a more critical perspective. The present book is also based on this important work, which has had considerable influence on sociolinguistic studies in globalized linguistic practices. There, it has specifically been taken into account as a far-reaching alternative to the often insufficient approaches in applied linguistics, where the notion of 'variety' erases the presences of colonial continuities and increasing social inequalities (Piller, 2016).

Besides the ubiquitous presence of English and other colonial languages, the knowledge of a foreign language is conceived as being *the* means of getting along while on holiday. In her study on how to acquire knowledge of holiday languages within a relatively short period of time, Alison Phipps (2007) focuses on the use of language as snippets and improvisation in tourism encounters, which are quick encounters where diverse people are brought together for a period of time and where a lot of information is exchanged in spite of the presumed insufficiencies of the linguistic skills of the players. Again, the notion of the linguistic standard and of normative ideologies about linguistic ownership, adequacy and mother tongue (Bonfiglio, 2010) are contrasted with the transgressive practices and self-authorship of tourists and hosts in utterly banal short-time migration contexts.

The subversive quick encounter plays a role in many sociolinguistic studies on multilingualism in tourism, as in an influential volume edited by Adam Jaworski and Annette Pritchard (2005), in which a decidedly interdisciplinary approach to discourse and communication in tourism is proposed, opening up a field of critical and small-scale studies rather than conventional overview work. Now taken as a kind of starting point for research on language in tourism, the problems and questions developed in this volume triggered a relatively rich literature on tourism and language practices in settings that had in common their embeddedness in imperial power regimes, their conceptualization as trivial consumerist experiences, and the powerful performance of enduring inequalities of players from the metropole and from the margins. The work by Crispin Thurlow and Adam Jaworski (e.g. 2010a, 2010b, 2011) remains crucial in contributing to this debate notions of class performance in language and tourism, entitlement and commodification, among others. By showing how speakers use and transform language in order to deal with and profit from growing mobilities, they open up an entirely new space for linguistic enquiry which deals with the material and economic values of languages. Besides that of Thurlow and Jaworski, a variety of work on linguistic objectification and commodification strategies in tourism has shed additional light on these processes (e.g. Heller *et al.*, 2014b).

The valuing of languages is also reflected in language ideologies, which are increasingly becoming an important field of research outside the more established contexts of anthropological linguistics and the ethnography of speaking. In tourism they turn, in a materialized form, into souvenirs and collectibles or, as Blommaert (2008a) suggests, into linguistic artefacts. Various case studies on language printed or inscribed onto objects (such as T-shirts, cups, fridge magnets, etc.) have argued that the interplay of text, material, shape and placement offers different readings that provide insights into very particular linguistic strategies, including irony, humour, hostility and creative play, turning marginalization into a considered and mocked condition and challenging hegemonic practices and ideologies (e.g. Coupland, 2010; Nassenstein, 2016). In self-reflexive studies on interactions with tour guides and street vendors, both Nico Nassenstein (2017) and

Edgar Schneider (2016), among others, add to this picture a detailed analysis of the use of language acquired in grassroots contexts as a working tool, which produces artefacts of another kind: jokes, slogans and sayings. While contributions on tourist language practices in the Global South remain underrepresented, interest in work that takes southern discourse seriously is increasing (Etieyibo, 2018; Nyamnjoh, 2017; Ochieng'-Odhiambo, 2013, among others). In a volume co-edited by the present authors (Mietzner & Storch, 2019), contributions that ask about language practices in postcolonial tourism settings are on offer – reflections on language and tourism discourse in Jamaica, Kenya, the Philippines, Iran and Morocco.

In another vein, the sociolinguistic study of tourism has benefitted greatly from a more literary and historical approach, for example in a contribution by Marcela Knapp and Frauke Wiegand (2014), who explore the physical and psychological challenges and limitations of the cultural immersion of tourists visiting African countries, which elicit specific performances and trigger particular forms of mobility. These performances and mobilities have in common with their representations that they are based on colonially embedded discourses of civilization, authenticity and culture, objectivity and post-colonial responsibility. This and other contributions argue that, while Africa is again and again advertised and performed as 'wild' and 'distant', travelling there is not about using or creating knowledge about this Otherness; instead, self-awareness and notions such as feelings and emotions are the focus of the experience. This is a salient part of development tourism, which goes hand in hand with other forms of tourism in most African holiday destinations. Many trips to African tourism destinations trigger a 'wave of helpfulness' that is hardly ever unproblematic for the recipient communities, who are expected to integrate into European ideas of performance and gratitude (Salazar, 2004). Aid tourism, which does not leave tourists untouched emotionally and is intended to offer ways of annihilating inequality through emotional investment as well as donations (Mietzner, 2019), forms part of what Binyavanga Wainaina (2005) has ironically and critically described as 'how to write [and think] about Africa'. Tourism and the encounters within this context trigger an immense industry of remembrance and worship, resulting in a picture of the continent that is reflected in a huge corpus of literature and social media texts. Wainaina holds up a shameful mirror to the reader of literature about Africa, showing that the discourse on Africa as an impoverished continent with a pitiable economy will be an ongoing problem, as discourse on Africa can only be performed in an economically profitable way at this level. The ruinous effects of imperialism continue to have an impact not only on the natural environment and built landscape at the sites of contemporary consumerist mass mobilities, but also directly on people and their social and cultural practices, including language. What remains is the urgently felt need for a linguistics of hospitality that takes encountering seriously.

Sound and Light Show

How does our interaction with sociolinguistic knowledge translate into our experiences in a touristified world? A morning in Mombasa, Kenya. We have taken the new ferry across the creek and reached, after circling around a number of *vipilefti* (the Swahili word for a roundabout), Fort Jesus, which is now a UNESCO-listed world heritage site. Mzee Mohamed, who works as a freelance guide, awaits us at the gate and gives us a tour: fort, museum, old town and market.

The structure of the fortress dates back to the 16th century and the Portuguese conquest (Sarmento, 2016). Today, Fort Jesus is a palimpsest. Atop the massive entrance gate, a Portuguese inscription says:

> Reinando em Portugal Phellipe de Austria o primeiro [...] por seu mandado [...] fortaleza de nome Jesus de Mombaca aomze dabril de 1593 [...] Visso Rei da India Mathias Dalboquerque [...] Matheus Mendes de Vasconcellos que pasou com armada e este porto [...] arquitecto mor da India Joao Bautista Cairato servindo de mestre das obras Gaspar Rodrigues.[1]

Nearby, mounted over the door to a courtyard in which a large poster advertises 'Home of the Best Egyptian Cuisine', there is an English inscription that reads:

BRITISH EAST AFRICA PROTECTORATE
PROCLAIMED 1ST JULY 1895
A.H. HARDINGE ESQU.C.3
COMMISSIONER

A map that illustrates the many trade routes that connect Mombasa with the world bears in its very centre the words:

المحيط الهندي
INDIAN OCEAN

Outside, graffiti on a coral stone wall states

KAGOD.

All this, we think, speaks for itself: different regimes of power, different languages, different writing styles and different places in the fortress.

Every single inscription, even that on the grave of an undated Christian on display in front of an Omani house (Figure 1.1), was intended for posterity. Nothing lasted. No tourist or pilgrim is there to read. It is just us and Mzee Mohamed. In one of the ancient chambers, a huge framed window, the glass still intact, stands leaning on the coral stone wall – the remains of a sound and light show that stopped operating five years ago. The glass reflects the gleaming sun. A few steps onwards, a vendor sells candy: *mabuyu, kashata, maembe*.[2] No chairs to sit on for customers. None of the queues that make visits to heritage sites elsewhere tedious. Undisturbed views from

Figure 1.1 Linguistic landscape referring to a pilgrim

the lookout to the old harbour. Three *majahazi*, as the dhows are called here. The water is too shallow for the big cruise ships and the tankers. The dhows go to Zanzibar with rice or corn, or come from there with other goods.

Walking down into the old town is quiet. No traders, who have given up waiting for tourists. No sound or light in the ancient houses. Being interested in culinary connections across the Indian Ocean, we ask for pastry makers and sweet shops, which had been many. They have all closed down; the economy has collapsed. Ruinous continuities (Stoler, 2013). Mzee Mohamed says that he feels cheated by the owners of the many hotels along the coast nearby. He says they tell their clients not to visit here. There are fears of robbery, crime, terrorism. There are smiles, tired workers and heat, we think, and some kind of resignation which is silent and unobtrusive.

Why should anybody book to go to a sound and light show, presented in at least four languages (Swahili, Arabic, Portuguese, English) in an old fortress that shimmers in red and gold and ochre, has neem trees and mangos and tamarind and frangipani growing in its ruined courts, has carved wooden doors wide open that stem from ancient times, and offers a cool breeze? Why should anybody go there, if being out after dark is considered dangerous?

Did times change from the better to the worse? When it was still considered safe to go out for dinner at the club, one could sit on a little green bench with a green sunshade mounted on top of it, which stood on wheels. The wheels ran on iron tracks which ran across town. Where there were no tracks, similar vehicles that were pushed or carried were available. Those who pushed or carried are those in the old photographs who stand at the sides, in coolie dress and without any names noted on the back of the picture. The fort survived well into the 20th century as a fairly intact structure. It was used as a prison by the British colonial regime. There is hardly any commemoration of this visible; only the plaques commemorating restoration efforts with European Union funds speak of some lasting attachment.

We ask about who was here before. Nobody but the Omani, says Mzee Mohamed. The Mijikenda-speaking communities who live in the coastal regions of Kenya do not play a role in this ruinous form of history making. History turns into something that is happening now: the angst of the tourists eats the soul of the old town, which is reduced to a fortress full of silence – an ancient site of plundering.

Sounds in a Village

Discourse on the emotional and intellectual gains of visits to sites of culture and history is ubiquitous. Aside from the production of texts by specialists, there is a rich repository of travel literature and texts created by tourists, particularly in social media.

In her blog on her travels and wanderlust, Viktoria Urbanek has written a post about her disappointment after a visit to a Maasai village in Tanzania. She wrote about feeling unwelcome, exploited and cheated by the villagers:

> Writing this shortly after it happened I still can't believe the rudeness of the Maasai and how it definitely changed my point of view about them. I had a totally different picture of them in mind and it seems like they have become slaves of the modern world and might think that money is the only way to go.[3]

Her otherwise average text triggered more than 200 comments, which do not seem to have stopped pouring in since the posting in 2015. While some commentators wrote that they were sympathetic, many others expressed their contempt:

> I hate that this blog trended enough to show up on a major site. I hate that I clicked on your blog post so that I added to your count of readers. I do know that I will NOT be following your adventures because you have given me a sour taste in my mouth. I would hate to stereotype others who hail from the same country in Europe that you hail from. I would hate to think negatively of a whole group of people from this one negative 'experience'.[4]

Other commentators expressed their readiness to help and take the distressed author to better places:

> Our advice to you and other travelers is to do research before you go somewhere! […] You can visit Hadzable or Datoga in Lake Eyasi which give you real good experience of local people (activities and way of living)!![5]

The globe turns out to be a giant shopping centre, it seems, where the goods people are searching for can be found if only they look on the right shelves: authenticity, groundedness, seclusion, self-actualization, discovery, independence and adventure. Yet it does not work that way. Whatever we might buy in this global shopping mall will soon be discarded, while those whose lifestyles are sold appear to remain the colonized Others in the images that are on display.

The village tour, like the visit to the national park and the dance show at the beach resort, is a canonical part of the attractions offered to package tourists in East Africa. The emotionally charged reactions to Urbanek's visit and blog entry reflect not only complex practices that are part of the continuing colonial ruination, but also make the contradictory construction itself visible: the tourists' desire for an experience of 'reality' is met with a request by the Maasai for an appropriate payment for the show that has been offered. Underneath the frustrated discourse on the entitlement of the paying tourist, there lurks something that is perhaps far more difficult to behold, namely the quest for absolution (Ruf, 2007). If only the inhabitants of the village, as the owners of 'African culture', could truly embody the antagonism of modernity, or contemporaneousness, which is tradition and allochrony (Fabian, 1983), then the imperial world order in which such visits take place (and on which they are based) could be perceived as unavoidable and true. But in the village, business is created for contemporary people, who seek to make some kind of profit out of one of the most difficult and destructive economies on the globe: tourism. The desire for absolution is not to be fulfilled here, and there is no such thing as salvation that comes out of tourism. In these fundamentally unequal encounters, there never is fairness in a deal, nor is there any healing on offer.

There certainly is, however, self-authorship, which allows both villagers and visitors to script roles in which they feel comfortable. During the many journeys on which the analyses presented in this book are based, we have taken part in village tours and related activities on offer for tourists. Even though the community (as a counterpart of the tourist resort) was already scripted as a site of allochrony, the village tour was often constructed and consumed as a hospitable place where the visitor could learn about cultural concepts of hospitality. The offer of food (Mietzner, forthcoming), shared space and activities (such as dancing or cooking) and the conversations among hosts and guests in a calm village atmosphere were educational and integrative in many ways. There were many cracks in the surface that allowed for subversion and encounters in a decidedly messy arena: in a village close to the beach resorts of Diani in southern Kenya, a small community of farmers had redecorated their village square so that paying visitors could be seated, offered refreshments and music shows and taken on a small village tour. The simple houses bore inviting inscriptions that said *HOME BREEZE, FARM, STABLE*, and the music and dance troupe had gradually professionalized their performance so that by the time of our visit they already had considerable experience in tourism performance through engagements in the nearby hotels. The ambiance was kept unobtrusive for the tourists, who were given much time to sit and relax. Yet all this followed a plan that emerged not so much out of any 'authentic culture' or 'true African village lifestyle', but out of the community's reflection on what the tourism business does to and for local communities in this region – barely anything. The village elder, who spoke various tourist languages including German, had spent many years at the beach trying to sell carvings or tours to people there. In the end, the children still had no shoes, and sending them to school was an impossible challenge. As a consequence, the community decided they would turn back to farming and to the village, and as the fields are poor and small, an extra income had to be generated. It did not suffice for any luxury, not even repair work on the houses, but it was probably better than work on the beach.

The extreme inequalities in such settings continue to be reinforced and affirmed by the global tourism industry, at the expense of the local communities, their environment and the integrity of the relationships people are able to share in these settings. Staged authenticity (MacCannell, 1973) and representations of invented traditions are resolved in spaces which are and are not directly correlated with tourism, such as in the homes of tourism workers (Meiu, 2017).

Work Trips

The notion of work is crucial for our conceptualization of the relationship between language and tourism in this book. While in many African communities language is conceived not as being separate from other languages but rather as fluid practice (e.g. Lüpke & Storch, 2013; Storch, 2016), in postcolonial tourism settings language becomes objectified. And it is work: learning the codes one needs in order to do business with tourists or other people on a beach, in bars or in the street, getting conversations started, attracting clients and convincing buyers. While language as part of the social repertoire in many Indigenous societies is both a cultural tool that helps to organize knowledge (e.g. in contexts of secrecy and initiation) and a social practice that resonates with notions of a community as being potentially open (Faraclas, 2012; Nyamnjoh, 2017; Storch, 2020), it is often presented and used in global tourism as WORK.

The different approaches to the contexts and practices of LANGUAGE-AS-WORK that are presented in this book have their starting point in a conversation on what would be a more appropriate way of describing a language today: not the conventional format of the grammar any more, but some other kind of approach that would make transparent the interrelatedness of different ways of speaking in a local repertoire, or the entangled world in which language exists. One of the most neglected notions of the value of language as such is work, the materiality of language and its political economy, which tends to be erased from descriptive approaches in linguistics. What if one decided to write about the 'grammar of labour' instead? How would notions of power, entitlement, ownership, relationships between people, community, etc., be expressed in such a description? Would there still be something like a chapter on 'the verbal system'? Or would there be an analysis of where to use imperatives and where not to, for example? There would, perhaps, be a description of the different ways of constructing imperatives in Swahili, where there are differences between the variety of language used among members of a village community in coastal East Africa and the language used among expatriates and their staff. Of course, there would be many insights to gain, and much to learn about the colonial continuities in linguistics itself.

Yet, the more we thought about work and language, or work and grammar, in the settings and places we were interested in or had some experience with, the more we saw the ubiquity of the global tourism economy. The work of an artist in Masindi (Uganda) or the work invested in the repair of village homes in Makhindhriso (Kenya), as well as the labour of women in a handicraft project in Jambiani (Tanzania), was impossible to dislocate from the ubiquitous presences of tourism: ecotourism, pro-poor tourism, voluntourism, package tourism, safari tourism, and so on. What was the grammar of work in such settings and how did it include the built environment of speakers, the mobilities of participants and the circulation of capital?

Furthermore, new linguistic roles appeared in our professional vocabularies – not just subject and object, agent and patient, but tourist and host. And we realized that we are tourists even where we are not, where we wish to present and construct ourselves as 'experts', 'researchers' and, well, 'linguists'. Would we then do tourism work as well, perhaps? What about the pocket guides to language often written by linguists? Isn't this tourism work after all, as these books are almost exclusively used by tourists?

Turning into a participant in the tourism economy can take place at any moment, and we might shed the role ascribed to us just as quickly: in and out of the arena in which language and social roles are situational concepts and practices. In the postcolonial contexts in which our work as linguists (specializing in African languages) takes place, we are constantly asked to reflect on the power inequalities and work practices based on colonial continuities. These reflections, we think, need to involve thinking about the impact of the political economies of tourism, the industry that overshadows many of our encounters, and the ways in which our mobilities are organized – travel to conferences, field trips, and so on. The effect of these colonial continuities that is most obvious to us is the ongoing ruination that maintains imperial order and capitalist extraction. This is the reason why we did not write about the 'language of work in tourism' or something like that, but simply about what we call *tourination*: tourism and ruination.

Yet *tourination* as a topic or field of linguistic work proved to be ruinous itself – a ruinous system, which ruins expert prestige within the discipline, where the presence of linguists at tourism sites is at best mocked: Doesn't the Africanist belong to the 'African village' in order to do 'field work'? We looked weird in our bathing costumes as we took notes, and our office spaces turned into a decoration hell as they suddenly had to house all the objects that had now become a scientific collection: mottoed T-shirts, fridge magnets, wall hangings. And we realized more than before that 'field work' is always situated within long-established power constellations, as are data production, theory making and the dissemination of scholarly work (Connell, 2007). Yet this is what tends to be denied in our academic discipline, where 'feelgood epistemes' (C.M. Bongartz, personal communication) remain firmly in place. However, we suggest that making inescapable role ascriptions and the illusion of objective observation in the most banal of research settings (tourism) more transparent in linguistic texts should always be taken into more serious consideration.

We are interested in three strategies of writing and organizing our text: (1) *Deformation* of the scholarly text, by changing genres and styles, contrasting and even mixing scientific writing and guidebook advertisements. The decomposed text, still based on established academic work, cites our existences as tourists and the privileges of 'field research' (Calacci, 2020). (2) *De-dissociating* the expert by not (simply) referring to sources, theories, thinkers and data, but bringing our own presences into the context, into the subject; becoming Others ourselves, thereby stripping ourselves of our academic certainties. (3) *Decanonization* of our reading habits, as we divert from the trodden path and read the apocryphal in great detail.

Structure of this Book

The chapters in this book are pieces of a mosaic that has to remain incomplete. They may all stand alone, as parts and pieces of a larger picture. The reader may begin with Chapter 2, which is on the historicity of the construction of the beach and of 'landscape', and explores how other ways of healing and of conceptualizing this space are always also there – the spirits of the past and present. But the reader might just as well begin with Chapter 3 on the philosophy of *Hakuna Matata*, which shows the interweavings and connections displayed by a journey that is already shaped by stereotypes and artificially created ideas. A trip to Kenya, which in terms of tourism is one of the most intensely constructed and discussed countries in Africa, full of ideas of the Maasai and *Hakuna Matata*, harbours profound philosophies and insights for travellers. A similar density of clichés, as well as destructive narratives on constructed African worlds of tradition and Otherness, is found where culture lurks: Chapter 4 deals with Blixen's texts on Africa and their afterlife.

Chapter 5 is a composition of two impressions of the first few hours after arrival at Mombasa airport. First experiences, which are actually the most formative experiences of a journey, exceed in their intensity everything that is to follow. On the way from the airport to the resort, one cannot ignore the linguistic ruins, as well as the visible dirt and the destruction on the roadside. In Chapter 6, the ruins that seem uncanny at sites such as the beach are explored in various ways. Hotels gone bankrupt, people who present embarrassing sights: ruination as over-consumerism and as a story of environmental change ends in a walk back to where it started – a remote country and a metropolitan airport.

Different and yet entangled stories about relocation and the ways in which relationships can be constructed at sites of deportation and refuge are offered in Chapter 7. This chapter moves away from the beach, in accordance with the tour programme, and turns to the national park where safari trips are made. Not far from its gates are the villages of those who once lived on park lands. Here, languages are managed in ways that resemble the management of wildlife diversity: counted, cared for, curated.

A Language of Relations is a chapter that deals with both invisible and visible boundaries that prescribe precise rules of behaviour for participants in beach life in Kenya. Recognizing and understanding these limits is challenging and often triggers unwanted emotional reactions. The beach also has a dress code, a leisure one, and this is what Chapter 9 explores further: the souvenir T-shirt with its often unpleasant prints, the appropriation of the Other's language, the text that is consumed while frying one's body by the water. What role do African languages play here? The same question could be asked of other texts, namely movies on travel to African countries. Often, such movies are about love, and often such love is constructed as emerging from sex tourism contexts. In Chapter 10 we discuss language, connections and bodies in some of these movies and consider the historicity of the images on offer.

Chapter 11 is on how Indigenous languages are performed in the museum and how they are turned into props. Many of the lexical items that are presented in the explanatory texts and performances in the museum are about food. The immersion into food preparation and local cuisine at the holiday site is dealt with in Chapter 12. Here, we explore the various practices of culinary tourism that play a role at the margins of mass tourism in East Africa and beyond. The book ends with a glossary that is also a conclusion. Terms that fly around in the travel spaces and concepts that abound in the marketing of tourism open up new fields of enquiry.

Notes

(1) Reigning in Portugal Philip of Austria the first [...] by his warrant [...] fortress named Jesus of Mombaca in the month of April of 1593 [...] That King of India Mathias Dalboquerque [...] Matheus Mendes de Vasconcellos who passed with a navy at this port [...] the late architect of India Joao Bautista Cairato serving as master of works Gaspar Rodrigues (translation our own).
(2) Sweets made from baobab seeds, coconut and mango, respectively.
(3) See https://www.chronic-wanderlust.com/why-i-regret-visiting-a-masai-village-in-tanzania/ (accessed 12 August 2020, link deleted).
(4) *Just an Observation*, see https://www.chronic-wanderlust.com/why-i-regret-visiting-a-masai-village-in-tanzania/" \l "comment-4153 (accessed 12 August 2020, link deleted).
(5) *Safari in Tanzania*, see https://www.chronic-wanderlust.com/why-i-regret-visiting-a-masai-village-in-tanzania/" \l "comment-51822 (accessed 12 August 2020, link deleted).

HAPTER 2

ible magical ways of healing — sea, spa, and skin: 45 minutes of consumerism

Ever found a nest of toothbrushes on the beach?

Or is it teethbrushes?

Come in and find out.

2 Terrible Magical Ways of Healing

Angelika Took a Plane to Mombasa

By the time she sat down in her seat, she already felt a slight tension in her back. Luckily, she was able to make reservations for a massage with Mary, a very busy masseuse who lives and works by the beach, just south of Mombasa. Mary's massage parlour is located in a small business centre in a back street, and is largely, Angelika says, a mimetic interpretation of the parlours at the beach resorts. The copy of the copy, in other words, as the 'premium' parlours and luxury spas there tend to mimic the Asian or African stereotyped wellness aesthetics of metropolitan parlours and spas, which themselves are mimetic interpretations of anything that tends to be considered 'authentic' but that actually refers to essentialist images of the colonial. In this mimetic mess, Angelika received a massage that resembled something she first thought was totally unrelated to the semiotics that prevailed at this slightly shabby site of the globalized wellness industry: the movements of Mary's hands were inseparable from Mary's voice; a hum at first, that went on alongside with the intensity of the touch, and increasingly sounded like the beginning of a trance. A short time later, there were also sounds that came from Mary's intestinal regions, and then, a bit later again, all this culminated in burping noises, belch after belch. The sounds came closer, very close actually, when Mary treated Angelika's scalp and forehead, with all the stomach noise, intestinal gargling, humming and burping right next to her ears. Why was she doing this, Angelika asked. Well, she was about to tear all the bad substances out of her client's body, Mary explained, which took their seat in her own body and then were removed with a belch. And another one, and one more. Very sorry, but this was inevitable; this was not a simple massage, of course, but healing. She was a healer, not just a masseuse, didn't Angelika know that? More noise, more trance-like movements, and then it was over. But be careful, Mary said, the pain might become worse during the next days.

Salty Water Ends Somewhere

Epistemes of salt water are epistemes of the infinite, the uncountable mass of water that covers the planet, that divides continents, defines mainland and island, that cannot be used to quench our thirst and that therefore makes hostile, deadly borders and barriers. In the languages that we normally study, as linguists who specialize in Africa, water – no matter what kind of water – is categorized as uncountable materiality, a liquid, and a mass. In noun class languages, such as Platoid and Bantoid languages, there is usually a noun class that denotes and contains liquids and mass nouns such as 'water', 'saliva', 'milk', 'oil', and so on. The noun class tends to be marked by a prefix *ma-* or *mV-*, which is widespread and is usually referred to by people interested in Niger-Congo noun class systems as 'class 6(a)'. In Swahili, 'water' therefore is *maji*, 'saliva' *mate*, 'milk' *maziwa*, 'oil' *mafuta*. In some related languages, actually many of them, class 6(a) also contains abstract nouns, which in a way also denote some kind of infiniteness. Endless love, never-ending desire …

Salty water, however, does end somewhere, is swept up onto solid ground, where it forms little pools or simply dissolves into the sand. Salty water, in other words, ends at the beach. Of course, this is contradictory, as the beach exists only because there is the sea, the waves that bring all this sand and shells with them and leave them there. But at the same time, the waves break at the beach, they stop there, and so does the salty nature of the water. A few steps further away, there are coconut palms – ideally, at least. In our imagination, the finite infinity of the ocean is never rubble or cliffs, but sand (white sand) and the sound of coconut palm leaves swaying in the soft breeze. The sound has a smell, of coconut oil, suntan oil, massage parlour aroma oil and coconut curries, preferably with seafood such as lobster. Lobster and prawns have a particular smell, not of the sea but sweet and elegant (almost like amber and musk).

Our imaginary soundscapes and smellscapes of the postcolonial ocean are premiumscapes. At the beach they are not phantasies, it seems, but they have a mooring onto something real, such as the shadow cast by our bodies and the tactility of the skin. We could sit under an airy shade, on the patio of a restaurant, with the palm tree leaves casting a soft shade on our bare arms. After the lobster, there is a dessert, little spheres covered with coconut flakes. Afterwards coffee. Later, there is a massage, oil dripping from muscles, always this oil.

The viscosity is as meaningful as the smell and the sound: not quick and liquid as water itself, but thicker and stickier. Oil, this noun class 6(a) substance, appears to bridge the gap between the ocean and the beach. It is made from the seeds of coconut trees, which grow near the tropical littoral space. Its thickness comes with a lobstery smell: land and sea, finite infinity, until the contents of the bottle have been smeared all over our skins.

In their study on the semiotic landscapes of luxury and privilege, Crispin Thurlow and Adam Jaworski (2010b) analyse the ways in which a whole industry is prepared for our premium phantasies, enhances them, triggers desires for the exquisite and promises to fulfil them. In the spa where we are given our oily massages, it is quiet. One can even hear the sound of the oil dripping onto the mat on which one lies. Silence. Thurlow and Jaworski argue that it is precisely this silence that signifies luxury and premium class in the semiotics of tourism. In ads for resorts, with their images of never-ending beaches and infinite blue seas and equally infinite skies above them, there is hardly any writing to disturb the exclusiveness and, if there is, it is about absences: just you and me and nobody else. The entire creation of imagined premium landscapes is about the non-presences of all who inhabit those spaces. The authors describe a contradictory situation:

> There is in fact a striking complicity and circularity in the relationship between tourism and globalization. A core assumption underpinning the discourses of both tourism and globalization is that of the promise of the 'global village' and the transformational potential of encounters with the Other. (Thurlow & Jaworski, 2010b: 187)

Class performance in the premiumscape turns this encounter into a phantasy. In this sphere we are not visible: in these images of the dream beach we are present as beholders. And therefore, there are no encounters, but phantasies of them, utopian dreaming of communication. But even language – in these ads it appears in the form of writing – is merely about the promise of interaction, not interaction itself. Noisy and messy encounters, the 'quick', as Alison Phipps (2007) has called it, are not yet there. Only silence as a resonance:

> Throughout, the deployment of silence and space is coupled with a number of other markers of super-elite lifestyle (e.g. haute cuisine, butlers, spa treatments, luxury brands [...]); it is also accompanied by hyperbolic claims to the exclusivity of the experience on offer, by flattery, and by synthetic personalization [...] of the implied traveler – for example, tag-lines like 'An invitation to the most exclusive club on earth' [...]. (Thurlow & Jaworski, 2010b: 192)

These performative practices have a genealogy. They are in place because they fulfil specific, important needs of an industry and those depending on it and its promises and offers, but they do not derive from these needs and desires. The practices that are at the base of the construction of premiumscape phantasies correlate with the invention of the beach – itself a process that has involved a large variety of cultural, social and political knowledges and activities, such as architecture, poetry, philosophy, hospitality, painting, friendship, gazing, healing, and so on.

Alain Corbin, in his famous book on the *territoire du vide* (1988), suggests that the historical connections – the genealogy of practice – can be traced back to the transitional era that led from the Roman Republic to the Empire of Rome. Citing Cicero, Pliny the Younger, Seneca and others, Corbin reminds us of descriptions of mansions by the sea, overlooking cliffs and beaches, where hosts and their guests enjoyed times of leisure which were devoted to the refreshment of the mind, self-discovery and self-fulfilment – away from the politics of power and professional duties. The Latin word *otium* here denotes a concept that vaguely translates into 'vacation': getting away from it all, to the beach, staying in a luxurious yet simple environment, resting, walking along the beach, spending time with friends and family. By no means was this considered a waste of time or laziness. The super-rich, for whom the *otium* was reserved, performed class and its privileges at the beach, in an environment that was carefully landscaped in order to resemble idealized mythological settings. And here the other meaning of *otium* comes into play: 'variety'. Surrounded by a bucolic yet built and arranged environment, one listened to the sound of a spring, birds singing, a soft breeze in the trees, the waves rolling on the beach. At the beach, one felt the soft sand underneath one's feet, gazed at the sea, and later listened to poetry or wrote a letter, or collected some shells and pebbles. Philosophical discussions were followed by a swim and then perhaps a game.

The beach was the favoured place for spending this kind of leisure time, and the mansions built by wealthy Romans remained in place for centuries; even after they eventually became ruins, the settings continued to be popular. Byzantine villas would provide similar views and settings, and provided the scenery for similar play and performance. Moreover, the meanings associated with these spaces and buildings consisted in part of knowledge practices that themselves constituted a means to define or demarcate class boundaries; Corbin emphasizes that later European educated elites knew that the littoral space had been a site of meditation, creativity, lust and play reserved for people like them. And because the relevant canon of ancient literature was so firmly implemented in these later times – and is still in place – elite travel in provincial France by the end of the *ancien régime*, and in 18th century rural England, for example, were journeys into new sites of ancient practice. The rituals of hospitality and friendship that formed a crucial part of the Grand Tour among the gentry cannot be made sense of, Corbin argues, without considering their reference to ancient models. The quest for self-actualization, the interest in exploring liminal spaces and the search for new creative means to express interiority all refer to the ideals of ancient class performance. In other words, the littoral space during and after the era of Enlightenment was not a space of liberation from class boundaries, but one of their consolidation. And seaside resorts remained relatively exclusive to an elite public until the mid-1800s, as did the spas, away from the sea. Corbin argues that they resembled ancient models even in their relative simplicity: the social life of the gentry, with all its elaborate performance, was much easier and cheaper to realize in the seaside resorts and spas, where grandiloquent architecture, spectacular events and the audiences for all this were readily provided. One saved considerable amounts of money by avoiding having to organize all this in the stately home, and had the advantage of being able to combine various social activities, such as matchmaking, furthering one's career, enhancing one's social network, and so on.

The interesting thing here is the conflation of the littoral space and the spa. By replicating the *otium* at other leisurely places, namely the spas in the interior, where therapeutic activities were the main reasons for a visit, places of healing and places of playfulness seemed to merge. At the beach, treatments with mineral spring water – spa practice – were transformed into treatments with sea water. And the ideas about bodies at the beach changed: foregrounding the need for healing, as a reflection of the effects of leading an ordinary life, of being vulnerable to gout and to consumption, turned these bodies into the medicalized objects of a positivistic scientific medicine. Writing about encounters with the spectacular at these beaches, Virginia Richter provides us with a construction that seems as familiar today as the idea of the leisurely *otium*:

> For whales, the beach is a thanatotope: they come to the beach only to die, thus profoundly disturbing our vision of the beach as a site of regeneration. (Richter, 2015: 160)

We wonder whether the beached whales really differ much from Professor Aschenbach at the Lido in this scripted space of decay and death. Bodies at the beach, ruinous and moribund. And yet, in 1803, Georg Christoph Lichtenberg published a plea for the establishment of public seaside resorts in Germany, where recreation and recovery were on offer to all, across class boundaries and regional divisions. The prerequisite for an appropriate visit to the seaside is an appropriate interior condition. Only those who can appreciate variety will truly benefit from their stay at the beach:

> Der Anblick der Meereswogen, ihr Leuchten und das Rollen ihres Donners, der sich auch in den Sommermonaten zuweilen hören läßt, gegen welchen der hochgepriesene Rheinfall wohl ein bloßer Waschbecken-Tumult ist; die großen Phänomene Ebbe und Flut, deren Beobachtung immer beschäftigt ohne zu ermüden; die Betrachtung, daß die Welle, die jetzt hier meinen Fuß benetzt, ununterbrochen mit der zusammenhängt, die Otaheite und China bespült, und die große Heerstraße um die Welt ausmachen hilft; und der Gedanke, dieses sind die Gewässer, denen unsere bewohnte Erdkruste ihre Form zu danken hat, nunmehr von der Vorsehung an diese Grenzen zurück gerufen, – alles dieses, sage ich, wirkt auf den gefühlvollen Menschen mit einer Macht, mit der sich nichts in der Natur vergleichen läßt, als etwa der Anblick des gestirnten Himmels in einer heitren Winternacht. Man muß kommen und sehen und hören. (Lichtenberg, 1994 [1803]: 96)[1]

What an interesting play with faded images of the premiumscapes of the past and the connection to those spaces that would become elite destinations of the late colonial era and of our times. The salt water that wet Lichtenberg's bare feet then is the same water that surges against the coasts of Tahiti and China today.

Coconut Palm Trees Make a Pleasant Sight

They frame tropical (colonial) littorals that are devoid of those decorative ruins that form part of the coastal landscapes of the north. Just the beach and a few boats, a dhow perhaps, small houses or huts; no beached professors. The tropical beach is a totalitarian colonial image that speaks of the inability of those who constructed it to face the consequences of what has been done. Gastón Gordillo, in his book on *Rubble*, remarks:

> Berman wrote that one of the features that distinguishes the bourgeoisie as a class is that it 'cannot bear' to look into the moral, social, and physical 'abyss' created by its own destructiveness (1982, 100–101). The fetishization of ruins is one of the ways in which the rubble created by capitalist and imperial expansion, and thereby the abyss generated by their destruction of space, is deflected and disregarded. (Gordillo, 2014: 254)

And then, at the moment that Lichtenberg gazes at a northern beach and thinks of other beaches afar, southern beaches, he gazes at all this from above – gazing at space and creating a landscape of mythological meanings. This way, one does not need to fear cracks and the destructiveness of one's own class. Like in the images from advertisements for sites of contemporary luxury tourism, such as those analysed by Thurlow and Jaworski (2010b), we are removed from all this. In these exclusive spaces, we watch the sandy tropical beaches from high up, like divine beings in an ancient mythological play.

This radical annihilation of bourgeois destructiveness, of the medicalized body and of the vulnerability of the colonizing Self (this contradictory construction of the omnipotent colonizer dying of consumption) produced totalitarian projects, where humanity was negated almost completely. And on tropical beaches, coconut palms and coconut groves were the backdrop when, in 1902, August Engelhardt sailed to the island of Kabakon where he established the Order of the Sun, which lasted three years. Members of the community lived on sunlight and coconut in order to obtain eternal life. After having lived on coconut for years, Engelhard still seems to have considered his emaciated body, disfigured by hunger oedema, divine.[2]

The Truth Will Out

Anne spends time at the beach. An outrigger boat, a *ngalawa*, at a Kenyan beach. Its name has been written on the hull in irregular letters of white paint: *itafamika*, a shortened or rather, non-standard, form of *itafahamika*, 'it will be known' in Swahili. As Anne reads it, she can only think of what is underneath the shiny surface of the premiumscape. She wades into the shallow waters, under which the beach continues, and takes a picture with her smartphone in order to store this name for research, quick and nonintrusive. But not quick enough to be unnoticed: a young man hastily moves between her and the *ngalawa*: 'Make a picture of me and my boat!' She knows it's not his boat; he is not one of the fishermen but works as a Beach Boy, selling anything that can be sold – guided tours, talk, souvenirs, dope, himself, himself in a picture. Since tourists have been staying away from Kenya's beaches and most of the resorts have been closed or are only half-booked, there are not many customers to buy from him or from the others. She doesn't have money with her right now and declines. As she tries to move on in the water, he wades along: 'You don't like black people! You have no respect!', he says. 'And yourself', she asks, 'how about you?' He says, 'I respect you so much Mama, my Mama Africa, so much respect.' Does she want to walk to the reef and see an octopus? She knows the reef is dead, killed by the feet of thousands of tourists who walked on it in the past, and she knows that there are no octopuses any longer. His name, he says, is Alan, actually Ali, and in order to be on the safe side, it's Aladdin. She should remember him, Aladdin, who will make her stay unforgettable. Again, she tries to tell him that it's only boat names she was after, and that she is sorry. His replies are predictable: there are no tourists around, they are all afraid of Kenya now, after the elections and before the next elections. The government, the tribalistic attitudes of the hotel staff and owners, this terrible country. But, apart from the cursing at the beginning of the conversation, never a harsh word about the stingy tourists who are there, nor about the neocolonial relationship between those who wait at the beach and those who reside in the resort. Neediness should not elicit any mixed feelings; she should feel happy and safe. 'We are no cannibals. We don't bite', he says. White guilt is conveyed softly yet intensely, with hugs, kisses and flattery: 'Mama, Mama Africa, Princess, Lady.' He remembers her name – Anne – every time they meet. Not many people remember his. He ignores her questions about anything that is local, and talks about global things, about those who found the loves of their lives at this beach and took them home to Germany.

At this beach, men are Boys, and they have nothing to do and nothing to say. In the resort, staff – men working in the restaurants, barkeepers, watchmen – offer themselves in similar ways, for a walk on the beach, for an inspection of a room, for a night in a client's bed. *Wo ist der Papa? Du must nicht allein sein, ich mach es dir schön* 'Where is daddy? You don't need to be alone, I give you a good time', several people tell Anne in German, once they know she is from Germany. Other languages that form part of their repertoires and that are also learned from interactions with tourists are Polish and Italian. The offers are most certainly triggered by a constant demand. In the evening, her bed is decorated with frangipani flowers and a note written on the back of an old receipt: WEL COME & ENJOY YOUR HOLIDAY OMAR ROOM BOY (K A R I B U MAMA).

Next morning, one of the souvenir vendors at the beach shows her his guestbook, a torn exercise book filled with the recommendations of previous customers. It also contains a small wordlist in Swedish, German and Polish with translations into English and Swahili. Somebody has told him that the acronym MILF means 'heart'. *MILF – moyo*. Another person has written *hakuna matata* in elegant letters. No problem.

At the shop in the lobby of the resort, there are *Hakuna Matata* T-shirts on offer, as well as *Hakuna Matata* coconut oil suntan lotion and, in case problems should occur, *Hakuna Matata* sunburn relief. The *Nomad*, an upscale restaurant and resort down the road, advertises its annual *Oktoberfest* with *Pretzel*, *Weisswurst*, Rye bread, *Radi* and *Lebkuchenherz*. The ad features a couple dressed in Bavarian gear, the woman wearing a blonde wig to cover her African hair. *Nomad* also has a bar, which features a hand-painted ad for Kenyan beer: *TUSKER Helping white men dance since 1922.*

At the beach, little shacks have been built just next to the grounds of those resorts that are still operating. Many of the shacks are massage parlours, decorated with frangipani flowers and painted with the names of their owners (*Stella Massage Parlour*). There is a faint smell of coconut oil. The hotels have their own spas, and most of them, such as the *Urembo Hair & Beauty Salon*, the *Southern Palms Beauty Salon* and the *Swahili Spa*, offer special packages for weddings: nails, waxing, bridal make-up. In the premium hotels such as the *Swahili Beach*, the hotel spa and the guests' bathrooms form an imagined continuity. Everything is wellness. Rooms appear spacious, as does the entire complex, with its cascades of infinity pools (one after the other), hallways, airy patios, arches and domes. The architecture replicates imaginary oriental designs, the cinemascope of the 1950s. The rooms contain furniture that looks like film props, and are constructed almost like film sets: both bedroom and bathroom open to the side of the entrance door and are about the same size. One could do a film about some prince of Arabia here. In the bathroom, there are oriental-looking copper containers for the usual amenities, and plastic containers with products that promise coconut oil. A folder that contains hotel information explains all this:

> **Slice of Paradise.** Swahili Beach is much more than meets the eye. Culture and history provides its very soul. From the architecture and the aromas to the music, it teases the senses. Drawing you in.

Just outside, under a coconut palm, there is a table with an arrangement of freshly harvested coconuts. An old-fashioned looking slate bears an inscription in white chalk:

MADAFU

– Hydrates the body
– Less Fat and no cholestral
– Full of Natural sugar salts and vitamins
– Better than most energy drinks

@ KSh 100=

Madafu is another word that bears a noun class 6(a) prefix and denotes coconut water. Close by, people dressed in white garments that look like what coolies wear in old films clean the beach. In a patio above it, drinks are served. Sipping a *Swahili Fruit Cocktail*, *Swahili Shandy*, *Coconut Kiss*, *Dawa Sawa* ('pure medicine') or a *Hakunato* (word play with the *hakuna* of *hakuna matata* and *Shakerato*), one can watch others doing their menial work, as they would in a colonial setting that would fit into the decade these concepts refer to. One could also lunch and watch in the *Baobab Restaurant*, which offers *Swahili Coconut Infusion* and imported fish.

All this forms part of a larger variation of the same motifs, namely SEX, FRIENDSHIP and HOSPITALITY (abroad), POETRY (on the menu), FOOD and FEAST, and LOVE (wedding packages, bridal make-up). The images are simpler in the premium class, and more elaborate elsewhere, but the same motifs and tropes appear over and over again. Eric Jennings (2006), writing about spas and the French colonial empire, reminds us of the colonists' strategy of appropriating places that were already in use, such as hot springs, waterfalls and beaches in Madagascar and Réunion, for example. These places were quickly made part of the colonial project of 'civilizing', through bureaucratic control and the implementation of new forms of healing and treatment, alongside new diseases and demands. At the colonial beach, practices and constructions of the *otium* and the northern seaside resort were implemented as copies of banal practice, not into a previously void space but into a pre-existing landscape of healing and recreation. The Kenyan resorts and sites of coconut oil application are palimpsestic spaces, where suntan lotion is smeared over the spice oils and soap of the *hamam* and the substances that were and are in use in those spiritual practices that also exist at the beach. There is something uncanny about the neocolonial seaside, something that emphasizes the purely imaginary nature of all encounters that take place there, instead of rendering them more real.

What might this be, though? One wonders what kinds of voices can be heard at night, when the resorts turn from places of embodiment into places of play. Where Anne stays, there is a weekly schedule of amusements. Always the same: Friday is always *ANIMATION CABARET NITE*, Saturday has the *HAKUNA MATATA BAND* playing, Sunday is for the *SENGENYA DANCERS* and Monday is *BINGO/TAMBOLA*. She really hesitates to take part, but is forced to by the research team. They claim a table with four plastic chairs, order white wine, smoke, and watch the other guests looking at their smartphones or at each other. The bingo game is wonderful. The animator sits on the small stage in front of them and announces each number that emanates from the plastic bucket that contains the little balls in a bingo code. Anne has never played bingo before and is smitten. What a wonderful person the animator is!

1	'Father George'
7	'walking stick'
11	'Mombasa-Nairobi highway'
8	'fat lady'
22	'two swimming ducks'
90	'top of the house'

Anne has all the numbers and wins the first prize, a wooden carving of a giraffe that is also an ashtray. She decides never to quit smoking. As she gets up in order to receive her prize from the hands of the animator, she also shakes hands with his assistant. The man who picks the balls from the bucket is *Baba Zar* – what a name! He who brings luck is called Master of the Zar cult, summoner of the spirits, healer and mediator, the one who talks to ancestral winds, boldly looks into the eye of the evil, understands even what the ghosts of Others wish to tell.

Enthusiastically, Anne asks her frequent conversation partner, Bwana Kiboko, about Zar at this beach. As they move out in his outrigger boat to the dead reef the very next morning, he tells her that the beach is inhabited by a huge community of spirits. They reside under the water, rather close to where she is now sailing. Sometimes they enter a boat. There are charms attached to most boats in order to prevent evil. He doesn't wish to say more, as this is delicate. Many spirits come to Diani beach with the Zanzibari fishermen, who travel up here with the fish and are traditionally hosted by the village for weeks. Magic transforms people's relationships, he says, and leaves them no choice but to be hosts and guests.

Language Itself Is a Curse

At the beach, whatever is said is said in a manner that defines and categorizes, describes and fixes. Men are available, stand there and wait. They are known by nicknames – *007, Aladdin, Kiboko, Papa Nikolaus* – or by the name of their trade. Women are married there, or at least are prepared for some kind of bridal existence, through spa packages and special offers of sex and companionship. Guests need massages and suntan oil, and are constantly anointed and basted with coconut oil, like ancient gods and goddesses whose statues received offerings of oil and milk. Fritz Kramer (1987) wrote that through a complex process of experiences of dispossession, enslavement and threats of the loss of spiritual sovereignty, in some African societies the temple was incorporated into the body of its individual worshippers. The destruction of shrines and religious objects and sites was compensated for, he thought, by turning the body into a religious object. Spirit possession and trance rituals could therefore be read as highly abstract and minimalist representations of religious mooring. The inclusion of the premium body, the tourist's body, in the religious economy of the Zar at a beach where its protagonists reside, also turns the ideology of superiority, power and control into mere bodies. And these bodies are oiled and prepared for their marriage to the spiritual powers that are present at these old beaches. This is nearly the most radical creation of power we are able to think of.

Although its spiritual meanings are secret, the language of the beach denounces its powers through its constant non-discursive nature. Language at the beach is not so much reaching out to the other but is intended to position: naming, placing, labelling. The elicitation of luxury brands on the signboards in front of ramshackle souvenir shops – *Gucci, Prada, Nike* – and the ubiquitous display of *Hakuna Matata* objects create a magical space in which everything is ready for the feast. Those who do not participate in the feast are already named and placed: *Kenya Kimbo*, expats who lack money and means and thus live on food prepared with the cheapest cooking oil[3] (Perbix, 2020). The names of those who do participate in the feast remain opaque. These prophetic signals are secret. There are other signs that are more accessible. The *Diani Beach Art Gallery* exhibits a sculpture by the Kenyan artist Gakunju Kaigwa which is called *Mami Wata*. An Airbnb guesthouse by the beach can be booked under the name *Mami Wata House*. A designer of surfing gear and beach clothing who is based in South Africa markets his products under the label *Mami Wata*. Each product comes with a fancy label with an image of a half-peeled banana and the following explanation:

> MAMI WATA IS THE AFRICAN WATER SPIRIT. IT IS SAID THAT THOSE WHO SHE TAKES TO BE HER LOVERS RETURN WITH A NEW SPIRIT AND BECOME MORE SUCCESSFUL AND GOOD LOOKING. COME WITH US. WWW.MAMIWATA.SURF

Attached to the label is a little card that promises 'AFRICAN NATURE CURES':

> DRINKING IS SURF TRIP MUTI TO CALL THE WAVES FOR THE MORNING. HUNGOVER? DRINK A STRONG BLACK COFFEE, A GLASS OF COCONUT WATER AND EAT A BANANA. THEN SURF.

Mami Wata as a brand; spiritual power as art, hotel and dress. Like Zar, it doesn't seem to matter where the spirit's cult was originally practised (in West Africa). Commodified spirituality transcends the colonial fixation of ethnicity, continent and the state.

Yet spirits do have their worldly existence, resembling humans in terms of everyday habits, livelihoods, preferences and desires. Kjersti Larsen (2008), among others, analyses spirit possession not simply as subversive ethnographic performance 'from below', but also as a concept of intimacy. At the beach of Jambiani on Zanzibar's east coast, several inhabitants of the village have seen the spirits, who live in a village that looks like Jambiani itself but is submerged under the sea. The spirits were spotted around nightfall – a blue hour phenomenon – and were easily identified, as they usually appear in halves: just one leg, one arm, sometimes a reddish eye. Spirits tend to visit humans by night, and it does happen that they fall in love with each other. There are married women who are visited by their spirit lovers every now and then, so that they have certain times when they do not sleep in the same room as their husbands. Other women cannot marry human husbands at all because they already belong to a spirit. Spirit marriages, the healer (*mganga*) Usi Juma says, are privileged unions, as they provide independence as well as social safety. Spirit lovers are premium lovers who provide personalized care and talk.

Language is not difficult, or special, in the spirit realm because it is immersive. Spirits speak every language possible and are able to interact with any person. For the language of the beach, terminologies such as 'multilingual', '(super)diverse' or even 'creative' do not seem helpful. It is kaleidoscopic, an oily trace of blurred writing on a palimpsest, which has no beginning and no end: unmeasurable substance, embodied and kept opaque in its complex semiotic dimensions. Positivism takes a dive.

Coconut Oil

Hengwa mafuta ya nazi Ya kutumiwa nyumbani
Na mtu kiwa hawezi Kwa ukavu wa kitwani
Huitia kwa henezi Yaburudisha bongoni
Mbwa-na-fu kwetu pwani Tena ni mti azizi.
(Nabhany, 1985: xxii)

The skimmed oil of the coconut, for home use,
And if the person is sick, also for the dry scalp
Is applied slowly, and refreshes the brains;
Is of profit in our coastal land, and is a beloved tree.

Notes

(1) The sight of the sea waves, their glow and the rolling of their thunder, which can be heard even in the summer months at times, against which the highly praised Rhine Falls is probably a mere washbasin tumult; the great phenomena of ebb and flow, the observation of which is always preoccupying without tiring; the observation that the wave which is now wetting my foot here is uninterruptedly connected with that which washes over Otaheite and China, and helps to make up the great military road around the world; and the thought that these are the waters to which our inhabited earth's crust owes its shape, now called back to these borders by Providence, – all this, I say, affects the sensitive man with a power to which nothing in nature can be compared, such as the sight of the starry sky on a bright winter's night. One must come and see and hear (translation our own).

(2) In our time of coconut oil hair shampoos, Engelhardt's story has been rediscovered as a tale of the colonial abyss: a Robinson Crusoe narrative whose protagonist ends up not as a virtuous member of an ambitiously colonialist nation, but as a starved cannibal, an ultimate Other, in a world that falls apart, as in Christian Kracht's *Imperium*. We personally prefer a more naïve take on northern meanings of colonial southern coconuts. Is there any cooler example of Europeans using coconuts than King Arthur's prancing horseless knights with their clacking coconut shells in *Monty Python and the Holy Grail*?

(3) See https://www.standardmedia.co.ke/entertainment/nainotepad/2001245702/10-ways-to-know-you-re-dealing-with-mzungu-amesota (accessed 12 August 2020).

CHAPTER 3

The Philosophy of

Hakuna Matata

3 The Philosophy of *Hakuna Matata*

> Africans share freely and enjoy smiles
> with everyone including visitors.
> It comes natural
> to the African and does a lot of magic and healing.
> That is why
> most people know the African smile to be infectious.
> Show More[1]

It Is Not Easy to Leave the Beach

The sun shines, and the soft tune that emanates from a small bar has a narcotic effect. A silly smile hides uneasiness, and turning around on the towel once more, from back to front, hides the impossibility of filling this non-existence with meaning. But does it matter, after all?

Welcome to the world of *Hakuna Matata*, a world without problems, poverty, borders or crime. A world that exists on the African continent. A world that can be defined as a country, which even has a language – called *African* by people who live outside – and where people often smile, even though their lives are harsh. Still on the sand, at the beach, with a travel guide printed on three pages which we found in the hotel lobby – FREE: TAKE ONE. Africa needs to be introduced to this constructed world, in order to show its visitors the simple ways to its heart and soul. This is not a new idea, by the way. Prior to the tourism ad that promised something in that vein, more than a century ago the first linguists to write about their disciplines were already claiming that it would be through the study of language that one could come to an understanding of how Africans think. People like Carl Meinhof and Diedrich Westermann were much referred to as founding figures of Africanistics, especially Westermann, for their ideas about how to understand Africans.

Language, African language, takes us there – where the heart is, where the narcotic substance of negated time does its work. Endless beach, endless slumber on this endless stretch of sand, lost in amnesia that frees nobody from the past.

And so everything remains in place, bodies on sand, language on paper. We learn how simple memorized phrases in a foreign language are accepted, by native speakers and non-native speakers, worldwide. The phrase *hakuna matata* is meant to express simplicity and represents Swahili as the language of the entire African continent; it stands for fun, adventure and romance.

The Genesis of *Hakuna Matata* as a Philosophy

Hakuna Matata is a catchphrase that originated in the 1970s as a reaction of the Kenyan government to turmoil in the bordering countries, reassuring refugees and others that Kenya was a safe country (Bruner, 2001: 892f). Ten years later, in 1980, the Kenyan music band *Them Mushrooms* wrote the song *Jambo Bwana* 'Hello Sir' and performed in tourist hotels along the coast of Kenya. Since the song was composed in order to entertain tourists, simple language was chosen for the lyrics: 'The message of *Jambo Bwana* is that tourists are welcome in Kenya, which is characterized as a beautiful country without problems' (Bruner, 2001: 893). The lyrics reveal nothing of the original context and rather read as a language course *en miniature*. The text resembles just any informal greeting:

> Jambo! Jambo Bwana, habari gani? Mzuri sana!
> Jambo, Jambo Bwana, habari gani?   M-zuri sana!
> 5.matter 5.matter 1.master 10.news INTERR.which 1-good much
> Hello! Hello Mister, how are you? Very well!

After *Them Mushrooms* wrote *Jambo Bwana*, the phrase *hakuna matata* became more associated with tourism (Bruner, 2001: 893), as can be seen from the extract of the lyrics where the phrase is extremely prominent. This was the intention of Teddy Kalanda, who composed the song in order to entertain tourists and to teach them some Swahili words.

> Teddy wasn't very sure about the song that he had only composed for purposes of 'teaching tourists Kiswahili' in Mombasa hotels. 'It was a kind of a joke for me,' he begins. 'I remember musicians in Mombasa would tease me and say it was a song for nursery school children. For me, as long as the tourists were enjoying and singing along with us and having a good time, we were okay. That was the idea of the song,' he says. [...] 'After performances in hotels in Mombasa as *Them Mushrooms*, I would sit at the bar where tourists would joke with the waiters as they tried to learn Kiswahili. The words they uttered the most were '*hakuna matata*,' and '*Jambo*.' So I got this idea that if I made a song with the simplest words in Kiswahili, they would learn the language as they sang along with us during performances.'[2]

Hearing the song in the hotels, from 1980 until now, has an interesting effect: *hakuna matata*, among other phrases from the song, like *jambo* 'hello' or *habari gani* 'how are you', becomes a part of the tourist's Swahili repertoire which, as Nassenstein (2016, 2019) states, is also taken back home as a linguistic souvenir.

When *Boney M.* released their song *Jambo Bwana* in 1983, the lyrics were in English and embraced all types of clichés about Africa. The song not only welcomed people to join the samba, a rhythm which is more naturally associated with Brazil than with any African country, but also displayed the stereotypical drum beat, singing, dancing and romancing which are used in all genres around tourism and media in representations of Africa.

In 1994, the Walt Disney production *The Lion King* finally turned *Hakuna Matata* into a phrase known and used globally, and furthermore into a philosophy accepted worldwide. The best-known song in the musical is *Hakuna Matata*, whose lyrics are even more simple and further removed from Africa, let alone Kenya: wonderful place, no worries, no problems, and all this is a philosophy (John & Rice, 1994).

As a consequence of the world-famous song from the musical, which was translated into 44 languages of the world (interestingly excluding Swahili), *Hakuna Matata* has developed into an expression for easy living or a life without sorrows or critical questions. Tourists traveling to Swahili-speaking countries are made familiar with this phrase and its imagined implications even before their travel.

That Walt Disney Productions picked *Hakuna Matata* may be due to the fact that it had – through the song *Jambo Bwana* of *Them Mushrooms* and *Boney M.* – already become known in East African travel contexts, and was already in use to express the problem-free philosophy. After the release of the musical, *Them Mushrooms* filed a lawsuit against Walt Disney Productions, claiming that the phrase was only so familiar because of their song and they wanted to have a portion of the income; they lost the case. Since then, *Hakuna Matata* has spread in publicity and can be found today in various contexts, such as in the names of hotels and bars all around the globe, as artwork for tattoos, as a mode of expressing 'easy living' in music and in many other domains. It imposes a picture of Africa on global audiences that does not permit any other perspective but that of amnesia.

Hakuna Matata Permeates Everything

In order to express feelings or attitudes, *Hakuna Matata* comes in handy as an appropriate metaphor for expressing a philosophy of life that would otherwise need many words. And so it jumps out at us, out of the lyrics of songs of all kinds, intended to make us feel relaxed and happy – emotions that are associated with vacation, a day off, leisure, time just for you and me. In the midst of life, while making a living and struggling with work, we are told to be carefree. An infantilizing way of conceiving the world, as a playground perhaps, with 'Africa' as an image of where this would be found: laughing people, always happy, the sun shining (Wainaina, 2005). Stereotyping the Other not just as an Other living in an Other time, but in an Other stage of life, not grown up but immature. This ostensibly positive view on 'Africa' is a harmful one which, however, comes packed as a feelgood leisure philosophy that fits into the lives of the urban conscious audiences who have the means to buy it – new elites in need of feelings. Living the hakuna matata philosophy and drinking green tea,[3] is what the German hip-hop artist *GReeeN* sings about in his song *Ich mach mich frei*, using keywords for a totalitarian grasp of the world, clad in fashionable discourse. While one drinks green tea and indulges in philosophy, there remains no problem to be considered at all, as whatever might remain left to explore is moved into the focus. Africa, via the *Hakuna Matata* catchphrase, again and again features as 'still unexplored', 'pristine' and 'authentic', as well as 'authenticating'. This phantasy of a far place that offers roots and realness is what Johannes Fabian (1983) has helped to reveal as a product of the manipulation of time. The Other is constructed as a kind of allochronic alter ego – us, but *then*, in the Stone Age, the Middle Ages, pre-modern times. Fabian argues that such a construction – the Other as an allochronic Other – is at the base of any discourse on development, any compartmentalization of the world into first, second and third, Northern, Southern, metropolitan and peripheral. A matter of power, what else?

The ever-relaxed and smiling African, this grotesque face of a human who seems to fail the earnest requirements of our time, is part of a whole repertoire of Othering strategies. Still living in a state of childhood, representing humankind's childhood on a large scale, this image forms part of a whole panopticon of demonized Others. It is no coincidence that this construction seems to work so well in music and the arts, such as film and literature. In her work on the *Colonial Art of Demonizing Others*, Esther Lezra (2014) uncovers the continuities of such constructions and the various tasks they fulfil. The demonizing of the Other, she writes, has not only helped to legitimize the colonial project, but also to imaginatively and discursively construct inverse configurations of power: '[...] the repression of Black suffering was psychologically transformed into emblems of white virtue' (Lezra, 2014: 120).

The construction of this asymmetrical relationship permeates the soundscapes for sale in the CD selections offered in the souvenir shops and the acoustic environment of bars and cafés. Even though tourist longings are usually addressed in texts that originate outside Africa, Kenyan artists also use such motifs to promote Kenya (or Africa) as the wonderful place that it is supposed to be. *Mani Productions Ltd.*, a producer and singer-songwriter from Kenya, sings about the light-heartedness of those who seek their salvation in the depths of a romanticized continent. Exotic Africa is where we are supposed to consume all that is refused to us where we come from; why else should we leave it? The imagined carelessness that surrounds Africa is everywhere – in narratives, songs, travelogues, movies, sun lotions (Figure 3.1). *Hakuna Matata* advises us on how to face problems, with an escapist imagination of the colonial place. Here, points of origin are, as Bruner (2001: 893) argues, made irrelevant 'and the distance is narrowed between us and them, subject and object, tourist and native [...]. What is left are dancing images, musical scapes, flowing across borders, [...] occupying new space in a constructed [...] border zone (Bruner, 1996; cf. Appadurai, 1991) that plays with culture, reinvents itself, takes old forms and gives them new and often surprising meanings'. Music as the key to emotions and expectations takes visitors into spaces of the unreal, enhancing their view of a stereotyped image of Africans and Africa, or even completely losing the connection to Africa as in the case of the hip-hopper Remoe, who titles his 2019 song *Hakuna Matata*, singing about sex, money and wealth without even mentioning Africa or the phrase in his lyrics.

While the music goes on, the use of *Hakuna Matata* as a label for hotels and bars all around the globe, as a motif for artwork, home

Figure 3.1 Sun protective lotion with the promise of no problem

decoration and for tattoos, as a mode of expressing 'easy living', has almost become tantamount to Africa itself. It is already sufficient to elicit the images needed to keep everything where it needs to be kept, roles neatly divided among the different players. The 'philosophy' and its oversimplification of Africa goes along with Swahili as the representative language of the African continent. Commodification of language snippets is one of the items on sale in the global tourism market. As *Hakuna Matata* is *the* typical slogan of easy living as the core of an imagined 'Africanness', it is easily employed in an ever recontextualized and commodified form in which it helps to create tourism mythologies (Thurlow & Jaworski, 2011). The snippets created from imagined language are part of globalization and the everyday ways in which particular social meanings and material effects are realized. *Hakuna Matata* is found on all possible items that can be carried back home to the tourist's place of origin; it is already there, on the bodies of those who travel, and taken to the holiday destinations where they intend to unwind. T-shirts, stickers and emblems, and body decorations of more lasting kinds, all speak of the blurred boundaries between Africa as a geographical place and an image. The occasional *Hakuna Matata* tattoo at the nape of a neck, on an ankle or a wrist speaks of a concept of Africa as a state of mind, a fantasy of a place where everything is the opposite of home – free, relaxed, sunny and happy, ever smiling. Africa's smiles and its *Hakuna Matata* language written under the skin turn into cannibalized Otherness, completely internalized Others, whose only meaning and function seem to be the provision of relaxation and leisure: the tropics as counter-space, a void into which all problems fall. Not simply a dump for the Global North's industrial garbage, but also for its psychological and ethical problems, in a Fanonian sense. A spell that erases the past, imperialism stitched onto T-shirts and etched onto the skin, where it is almost out of sight for those whose bodies it adorns.

And while we contemplate the tattoo, other things catch our attention.[4]

- A box of shisha tobacco, *Hakuna Matata* flavour, with a sticker that says 'smoking kills'.
- A bright yellow bottle of *Hakuna Matata* suntan oil that comes too late.
- A packet of *Hakuna Matata* spices for chicken.
- A bottle of *Hakuna Matata* lemon and ginger syringe for summer.
- A *Hakuna Matata* vinyl pack, for whatever purpose.
- A pair of *Hakuna Matata* sneakers, with *pole pole* written on the side.
- A *Hakuna Matata* wall art sticker set.
- A *Hakuna Matata* gift voucher.
- A six-pack of *Hakuna Matata* beer.
- An organic candle with *Hakuna Matata* written around it.
- A *Hakuna Matata* dog collar in blue.
- A *Hakuna Matata* T-shirt in ochre.

- A bottle of *Hakuna Matata* wine, 'with a hint of sweetness, some fruity flavour, and a bit of floral, [it] is fun drinking. Ideal for an evening spent relaxing on the balcony fully embracing this wonderful phrase'.[5]
- A *Hakuna Matata* tote bag for all the things we have bought.

Into the *Hakuna Matata* Void

Let us go then, to experience life, life without any problems but with itchy skin.[6] All this *Hakuna Matata* has made us curious about what is out there. In order to be really well prepared, we might first want to read a travelogue. A travelogue about a trip to Tanzania is headed by the phrase, 'How I found my Hakuna Matata',[7] and tells us about the travel pace (slow). In a travel blog, the question 'How do you say hakuna matata in Thai?'[8] is posted.

Language is turned into a snippet, a slogan, a catchphrase that is found on all the other things, too: not just something that emerges out of the encounter, but something that is consumed together with wine and wall art stickers. Travelling into the *Hakuna Matata* void is not a journey into the unknown and adventurous, it seems, but into the known and labelled. Translation appears to be easy: how do you say *hakuna matata* in your language? What language does it come from, after all? If there is a mooring that can be offered, then perhaps one that suggests *Hakuna Matata* being the 'spirit of Africa' and Swahili the 'language of Africa' would be helpful. It represents complete ignorance of African linguistic realities, of language practice as well as politics,[9] that a snippet is turned into a characterization – a great metonymic move in the discourse on others that is used to advertise almost anything here. And so Swahili is everywhere there is a need to represent oversimplified constructions of Africa: in Cape Town's tourist-geared *Victoria & Albert Waterfront*, there is no sign of souvenirs relating to any of the regional languages; instead, postcards and T-shirts with *Hakuna Matata* inscriptions are offered. In other parts of the continent, where international package tourism plays a similarly salient role, such as Rwanda, Uganda, Gambia and Senegal, *Hakuna Matata* lodges and bars abound. Airport art and souvenirs with *Hakuna Matata* inscriptions are available almost everywhere, while inscriptions that derive from other languages are as rare as they are in South Africa.

Here, *Hakuna Matata* is separated from Swahili to such an extent that it has developed into an overall concept in which borders play no role and in which 'Africa' is perceived as a single entity. This dissolution of countries and sociocultural groups results in an ahistorical void, an abyss of timelessness, a contradiction that denies any experience and encounter that may take place. In its globalized form, in which it is often reduced even to a symbol, *Hakuna Matata* has strong overtones of counter-culture, enhancing a concept of Africa as the desired 'Other'. 'Hakuna Matata tattoos are almost as popular as Lion King tattoos and many of the kids who grew up watching The Lion King and singing the popular song Hakuna Matata have a very deep attachment to the phrase', is what the website *TatooEasily*[10] gives as an explanation for the genesis of the popular *Hakuna Matata* tattoo. A symbol that looks more like the logo of a video game than something stereotypically 'African' often accompanies tattooed *hakuna matata* phrases, or replaces them altogether. The origin of the symbol is undocumented, but can be traced back at least to the South Korean film *200 Pounds Beauty* (Yong-Hwa, 2006), as Nassenstein and Rüsch (2019) reveal. K-pop artists such as *Walwari* use the symbol in order to refer to an urban hip lifestyle, which differs from what might be expected by older people of young South Koreans. A counter-image of a life spent working hard and being exhausted – cool, relaxed and a bit cheeky. *Hakuna Matata* constructions are constructions of imagined 'African' spaces in which people while away time and enjoy hedonistic lifestyles that are not affordable otherwise. Doing nothing, doing things slowly, never again taking the regime of appointments and schedules seriously.

Enjoying the wonderful phrase on a balcony of summertime fun.

Lighting an organic candle.

Staring at the wall which still is a wall. A wonderful phrase has been glued onto it.

Staring at a poster that advertises a festival at a German school[11] that offers courses on hospitality and tourism. It offers an evening of 'HAKUNA MATATA – The Spirit of Africa'. This includes presentations on 'the world of spices' (*Die Welt der Gewürze*), the tasting of 'African specialities' (*Afrikanische Spezialitäten*) and a 'drumming lesson' (*Ein Trommel-Workshop*). Below, very much in accordance with Fabian's (1983) assumption that any denial of co-evalness that involves constructing the Other as an allochronic Other serves as legitimation for development politics, the statement that 'not only do you get to know a country better, you also support an aid project in Africa by attending the event' (translation ours) follows.

Staring at a mural in Christiania, Copenhagen, that asks visitors to buy shares in order to support the neighbourhood (Figure 3.2). Freetown Christiania is a commune that is mostly conceived as a social experiment, helping to develop new ideas and practices of anarchistic and alternative ways of life and, for many, also providing an open space for a polyphonic critique of the state. The mural is not far from the entrance gate: flags representing countries, and text representing national languages: Norwegian, Danish, Swedish, English, German, and so on. A flag with the silhouette of Africa. The African flag. Africa is a country, and its language sounds familiar – Swahili, the language of Africa. *Saidie kuipata Christiania. Nunue cheti za uhuru hapa* 'Support Christiania. Buy a share here'.

Figure 3.2 Swahili as the language of Africa: A wall in Christiania (Copenhagen)

Swahili as the language of the African continent appears where there is a need for order on large scales. This is easy order – feelgood epistemes turned into matching sound. Language is made to sway here, softly moving through the fermenting heat. Language coagulates to sticky matter that covers all cracks and holes. The postcolonial Other appears to act in *slow motion*, uttering a slurred 'no problem'. A totalitarian view of the Other indeed – calm, relaxed, almost mute – and a perspective on Africa which is now a cornerstone of the tourism industry. 'Happiness' and 'contentment' are other terms that often appear to characterize its phantasies. No problems, don't worry; there is happiness and contentment under this southern sun. The lives of Others remain opaque in all this intimate intrusiveness (see also Appadurai, 1988: 46). They appear to be nothing more than a tourist event which, however, 'is presented as ethnography, and when performed in the round as a total immersion experience, [...] is the ultimate case of ethnographic realism. The authority of ethnography is used to authenticate a tourist production that shares its realist effects' (Bruner & Kirshenblatt-Gimblett, 1994: 458).

We can turn the whole thing around, though. Instead of sticky description there could be experience. Describing the *Hakuna Matata* void is one thing, but becoming part of it, being drawn into it through the tunes of these old songs, is another: being *Hakuna Matata* tourists.

This is us. Us, who went to Diani, the Kenyan beach, as package tourists. Us, who perceived the cordiality of the staff not as cordial, but as oppressive and authoritative. Us, who found ourselves in a state of being unable to be ourselves, in a state of being created, constructed and moulded, ending up in a permanent state of hopeless self-reflection. A strange setting, an all-inclusive hotel in which tourists from our country spend a short-term holiday of ten or so days. What we experienced were our own perceptions and feelings as scholars who for decades have been debating about colonialism, post-colonialism and Othering. And then we found ourselves in a position of constant self-reflection, oppression and returning soupçons of guilt, bad conscience and fitted friendliness.

At the airport, the smile of a Mombasa airport employee elicited a '*Jambo!*' from the experienced Kenya traveller, which was responded to by the employee with an even wider grin – '*Jambo! Hakuna matata!*' There you go. *Hakuna matata* is an established greeting in Kenya's tourism spaces (Figure 3.3).

The hotel itself was an enclosure that contained phantasies and imaginations. Imitations of elephant tusks, artificial jungle, pond and all. In every hotel there was a display of phantasies of Africa, different ones. Makuti roofs, white walls, black beams and a Swahili-Arab looking concierge with a fez on his head. The shop offered *Hakuna Matata* sun care products. Sun milk in our bags, our bags in our sweaty hands. At the pool area we were greeted by the animation team. It was their job to be in close contact with the tourists: '*Hakuna?*' – '*Matata!*' These were the conversations we shared with a large number of people.

Figure 3.3 Mombasa airport waving goodbye with best greetings

Sometimes, we tried refusing to give the answer, as we still thought that we were reflective academics who would like to retain their level of integrity, knowledge and self-esteem. A correct greeting in Swahili would be different (Yahya-Othman, 1995). But after three more '*Hakuna?*', which now sounded more like a threat than a question, we gave in on the conventions and politely replied '*Matata!*' – 'Problems!' This was one of various small steps that turned us into *Hakuna Matata* tourists, leaving us with less and less agency as we strove to embody the ease of life, seeking to reverse colonial hierarchies so that salvation would finally come as the allochronic Other took control of us.

The construction of the *Hakuna Matata* tourist makes sense for both hosts and customers, in order for both to gain as much from the whole thing as possible. The philosophy of *Hakuna Matata* almost catapults the tourists into a state of ease and happiness, which seems necessary for them in order to be able to accept the debilitating enclosure in which they are kept. Outside, we are told, there is crime and danger. The word for 'danger' in Swahili is *hatari*. *Hatari!* is the title of a very successful movie made at the end of Hollywood's golden era. John Wayne is in it.

Moving Out of the *Hakuna Matata* World

Them Mushrooms are considered a national icon by many Kenyans. Their history as a successful band has lasted for more than 45 years, and they remain very much present in the country's entertainment industry, having produced a large amount of both hits and film music. During the elections in 2017, they made their contribution, which was duly noted in the Kenyan news:

> Famed for re-recording big old songs, taking their many adoring music fans down memory lane, one of the country's most enduring bands, Them Mushrooms, is set to celebrate a milestone: its 45th anniversary in entertainment.
>
> This week, as the country prepared for the fresh presidential elections, The Mushrooms were not left behind. They enthusiastically threw in their lot with the rest of their compatriots, by doing a patriotic song Hatutaki Matata, a remix of their long-time international chart buster, Jambo Bwana.[12]

And suddenly, the 'tourist anthem', as the *Daily Nation* puts it in the same article, turns into a political tune: *hatutaki matata* 'we don't want any problems', asking the Kenyan public to contribute to peaceful and orderly elections. A 'patriotic song' indeed. It does not begin with the simplified *jambo*, but with *hamjambo*, which would be the appropriate greeting, and consists of more complex Swahili lyrics instead of the simplified tourist language on offer elsewhere.

While the original international release of *Jambo Bwana* had been advertised as 'the music that makes you HAPPY',[13] the band dealt more critically with it, as those who were made particularly happy might not have deserved it. *Hakuna matata* has not only become a phrase that can be modified in order to address political problems, but also one that elicits rather bitter feelings about exploitation.

Twenty years after the release of the Walt Disney production *The Lion King*, Disney Studios have trademarked the Swahili phrase *hakuna matata*. Regardless of whether they will actually keep this trademark or not, it is very difficult to understand how a phrase that has become a household word in African tourism is supposed to belong to a production company. And yet, this is not simply about language; *Hakuna Matata* has turned into a source of income for East Africans working in tourism, for people making fun balcony wine, for people producing wall art stickers. No laid-back swaying here any more – *Hakuna Matata* has to hurry, because otherwise payments to the production company are due. Any T-shirts, mugs or pendants that bear a *Hakuna Matata* inscription on them would have had to go through the complicated process of being registered. And the musicians, and their song, their patriotic song: what about this? A year after the news on *Them Mushrooms* and their contribution to the Kenyan elections, another news outlet wrote about them again, this time in California. The celebrity news site *TMZ* reports that band member John Katana is fighting for his and *Them Mushrooms'* intellectual property rights. The short announcement mentions 'colonialism and robbery', and theft:

> **Disney** didn't just commit cultural appropriation when it locked down the rights to 'Hakuna Matata,' the company actually stole the saying from an incredibly popular Kenyan band … this according to the group's frontman.[14]

Meanwhile, the online platform change.org has initiated a petition to sign against the Disney branding,[15] and comments on the insult of grabbing part of a people's language:

> While we respect Disney as an entertainment institution responsible for creating many of our childhood memories, the decision to trademark 'Hakuna Matata' is predicated purely on greed and is an insult not only to the spirit of the Swahili people but also, Africa as a whole. The movie is set in Africa and the characters have African names, which further makes the decision to implement the trademark a perplexing one. The term 'Hakuna Matata' is not a Disney creation hence not an infringement on intellectual or creative property, but an assault on the Swahili people and Africa as a whole. It sets a terrible precedent and sullies the very spirit of the term to begin with. At a time when divisiveness has taken over the world, one would think re-releasing a movie that celebrates the unlikely friendships, acceptance, and unity, Disney would make a decision that goes completely against these values.

As if the trademarking was the major problem here, all the Othering, exoticism and defacing of Africans though images of happiness and contentment in the post-colony are removed from the discussion. African words have been stolen and should be given back, in order to make things good again – like artefacts in European museums, like things. But this is more complex; this is about continuities of colonial thought. The catastrophic capture of language here happened a while before Walt Disney entered the scene.

Not that anybody would have noticed.

A subversive reading of *hakuna matata* is present in various pop-cultural forms, such as the work by the Mexican visual artist José Rodolfo Loaiza Ontiveros, who critically engages with Hollywood's hegemonic genres, and in particular with the gender ideology present in Walt Disney productions. His painting *Hakuna Matata*[16] is a parody of the MGM logo, an icon of the once most powerful studio in Hollywood, which produced films and images that greatly contributed to the spreading of powerful heteronormative ideas about sex and gender. We see a logo in which Leo, the company's heraldic animal, does not roar majestically any more, but licks his testicles, which directly face the viewer. Instead of *ARS GRATIA ARTIS*, the company's mocked coat of arms bears the inscription *PEACE AND LOVE*. The banner above the logo reads 'Don't Worry be Happy' (instead of 'Metro-Goldwyn-Mayer') and, to the left and right, instead of 'TRADE MARK', an ironically programmatic 'HAKUNA MATATA' can be seen. Two stereotyping texts on Black lifestyle are brought into a relationship here: Bobby McFerrin's song and Walt Disney's appropriation of *Them Mushrooms'* song. This parody engages critically not only with the problematic representation of the world in Hollywood films, but also with the intellectually debilitating use of the phrase *hakuna matata*, and its theft.

Worms Moving In

At Diani Beach and elsewhere on the southern Kenyan coast, the sea has begun to behave a little bit unpredictably. Currents have changed and children have drowned, where a year ago or so swimming was easy and safe. Fishing grounds are disappearing and reefs are turning into dead matter. This is not the effect of anything as abstract and distant as climate change. Offshore sand harvesting and the sand trade have become one of the most profitable businesses in the world. The removal of large quantities of sand changes currents and destroys ecosystems close to where it is extracted, while helping to build urban landscapes elsewhere. Sand at the bottom of the sea, as hard to imagine as language in the mouths of so many people: how can this come to be traded? But it is. And those who had it and lived with it do not get anything for it, as has always been the case throughout this violent colonial encounter – which we like to imagine as an encounter that began at the beach, by the way. After so much has been taken away, it is now the beach itself that seems to be going. Busy ships with nothing but sand in them, on their way to the construction sites of new supercities and hyperharbours.

This is not the first time ships have carried sand:

> A daily reminder of the Atlantic slave trade had an impact on our archaeological work in the Congo: the inadvertent introduction by slaving vessels of a species of sand tick known locally as a chique *(Tunga penetrans)*. When one walks on these infested beaches they burrow into the feet, especially under the toenails, where they rapidly grow into a painful, pea-sized creature that must be carefully dug out before blood poisoning or gangrene can set in […]. The creature arrived on the Congo shores in the sand ballast of slave ships as they returned to Africa lightly loaded from Central and South America. The chiques were deposited on the beach when the ballast was dumped in preparation for taking on a new cargo of slaves bound for the New World. The tick and its associated disease, *tungiasis,* has now made its way southward from the slave coasts of Central and West Africa to the Cape of Good Hope. It is now continuing northward up the East African coast. (Denbow, 2014: 153–155)

And as the ticks move on, a group of people have assembled to protest against the continuing exploitation of everything that surrounds them, of the sand and the sea. Their battle cry is *hakuna matata*. It sounds fierce.

Notes

(1) Based on https://www.youtube.com/watch?v=fHzRBlypnQo (accessed 12 August 2020).
(2) See https://nation.africa/kenya/news/30-years-of-hakuna-matata-well-almost--624554 (accessed 12 August 2020).
(3) Based on our translation of the German lyrics.
(4) This list is the result of several shopping trips, in shops and online.
(5) See https://winemoments.com/products/hakuna-matata-white (accessed 12 August 2020).
(6) *Hakuna Matata* after-sun lotion helps.
(7) See https://blog.africadreamsafaris.com/2017/05/15/how-i-found-my-hakuna-matata/ (accessed 12 August 2020).
(8) See https://urtravelogues.com/2011/08/01/how-do-you-say-hakuna-matata-in-thai/ (accessed 12 August 2020).
(9) Consider the various attempts to establish Swahili as a pan-African mode of communication. During the South African struggle for liberation from the Apartheid regime, many exiled activists and politicians learned Swahili and used it as a language that helped to establish networks with other liberation strugglers elsewhere (Gross, 2016). More recently, at his farewell speech as chairman of the African Union in 2004, Mozambique's then President Joaquim Chissano surprised other African leaders with his remarks in Swahili. At that time, the AU used only English, Portuguese, Arabic and French as official languages. Chissano said he spoke in Swahili to urge African nations to promote and use Indigenous African languages. Shortly thereafter, the AU adopted Swahili as an official language. See https://qz.com/996013/african-languages-should-be-at-the-center-of-educational-and-cultural-achievement/ (accessed 12 August 2020).

(10) See http://www.tattooeasily.com/30-amazing-hakuna-matata-tattoo-design-ideas/ (accessed 12 August 2020).
(11) See https://www.kks-marburg.de/index.php/schullebenkks/gastfreundschaft-und-ernaehrung-2/1174-hakuna-matata-ein-hauch-von-afrika (accessed 12 August 2020).
(12) See https://nation.africa/kenya/life-and-style/weekend/them-mushrooms-band-to-mark-45th-anniversary-this-year--469700 (accessed 12 August 2020).
(13) See https://www.discogs.com/de/Them-Mushrooms-Band-Jambo-Bwana/release/2401046 (accessed 12 August 2020).
(14) See https://www.tmz.com/2018/12/22/disney-hakuna-matata-controversy-john-katana-them-mushrooms-lion-king/ (accessed 12 August 2020).
(15) See https://www.change.org/p/the-walt-disney-company-get-disney-to-reverse-their-trademark-of-hakuna-matata (accessed 12 August 2020).
(16) See https://laluzdejesus.com/jose-rodolfo-loaiza-ontiveros-disasterland-chris-bales-anthony-purcell-richard-meyer-ave-rose-click-mort-d-w-marino-byung-heather-oshaughnessy/hakuna-matata/ (accessed 12 August 2020).

CHAPTER 4

Karen

4 Karen

A House

Each time I watch the movie *Out of Africa*, I get swept up in the sheer beauty of it all – the scenery, the sets, the music and the love story. It's hard to believe it was filmed in 1985 because all these years later, it feels timeless. I still love Karen Blixen's house just as much as I did the first time I saw it.[1]

Nairobi offers many interesting sights for movie site tourists. But one of these attractions outshines them all: the house in which Karen Blixen lived. It is a one-storey bungalow with a shady porch, large doors that allow fresh breezes to sweep into the living room, and private rooms overlooking the rolling Ngong Hills. The house was built in 1911 by the Swedish architect Åke Sjögren, whose masterpiece it has remained until today. This is partly due to the fact that nobody seems to remember any other house ever having been built by this man, and partly due to the fact that even if he had built another, nobody would care. For the sheer beauty, scenery and love that inhabit this house are, as writer Julia Sweeten points out, hard to exceed; this is not an ordinary house, it is a romantic house. A house built by an enigmatic architect, by the Swedish sphinx of architects, only to become the home of timeless love just five years or so after its completion. And it has remained this way, in spite of changing owners and inhabitants in later years.

The house is located 16 kilometres from Nairobi's city centre, and can be reached by taxi or bus along either the Ngong or Langata roads. The little journey out of town is one of the things that are considered highlights of a visit to Nairobi, offering impressions of an undying past. One can not only drive out there, but one can actually enter this house, this home of love and romance. Since 1986 the bungalow has contained a museum dedicated to Karen Blixen. Surrounding its spacious grounds, there is a nature trail into virgin forest as well as the surrounding gardens (over 100 plant species), and there are birds (116 species), butterflies and squirrels that can be watched. The museum is open daily from 8am to 5pm and charges KSH 200 (approx. 1,80 $/1,30 Pound) per person (residents)/ KSH 1000 (non-residents). It can be visited on guided tours only, which are offered throughout the day. The souvenir shop offers books by Karen Blixen, African beadwork and wood carvings, as well as DVDs of the movie that was inspired by one of her books. The guestbook contains many favourable comments, mostly by American and Danish visitors.

Julia Sweeten, who writes about houses, has explored the real Karen Blixen house in more detail. Another 'real' Karen Blixen house once stood nearby, and was the previous home of the widow of Jomo Kenyatta. It must have been overgrown by ivy and other plants, because Sweeten mentions that, in order to be turned into Blixen's 'house', it needed to be cleared of vegetation. After this, some architectural additions were made in order to make it more Åke Sjögren-like, and different vegetation had to be planted. Then furniture and fresh paint were added, making it a lovely house where, finally, the 'newly-wed Blixens' could move in. But matrimonial bliss didn't last long because Karen contracted syphilis from her husband, the 'Baron'. A pilot then flew by, dressed in white, washed her hair in a portable hair-washer, and they fell in love. This was true love, not syphilitic love, and Karen enjoyed some newly found freedom which she expressed by wearing complicated breeches instead of complicated gowns. She and the White pilot roamed through savannas and wildernesses, living rooms and porches. Unfortunately, this was not meant to last forever, because the pilot wanted to fly, and died in a terrible plane accident out there in the 'African vastness', where his remains lie until this day. x.

Sydney Pollack's film, based on Blixen's book, premiered in 1985 and remains well known and much loved, especially by European and American audiences. Sweeten's reminiscence of the film's setting, namely the home of its main character (the HOUSE), meticulously traces all the architectural modifications and preparations that were needed to create the ambience of colonial life that obviously enthused audiences immediately and continues to do so. And so, an old dairy turned into timeless scenery for an anachronistic story, Hollywood plastic, consumable and imitable.

While the actual Karen Blixen house stood nearby, a bit neglected and occupied by a governmental training facility and therefore not available as a site to make the film, its revenant became a powerful yet elusively immaterial icon of what audiences elsewhere expected of Africa. As the simulacrum of Blixen's house and life re-created and affirmed imaginations of colonial image politics, spaces and power relations, new interest in the writer emerged. Within a few months, a Danish woman, a member of a privileged social group of urban bourgeoisie and a settler colonist, was turned into a symbol of anti-bourgeois libertinage, the fight for women's rights and the subversion of colonial regimes of bodies and spaces. The beautiful style of her writing, her extraordinary appearance and her former fame as a mid-century New York intellectual idol provided sufficient means for such a popular resurrection. And soon the real house was resurrected as well. Mbogani, as it is called, was finally bought by the Danish state in the early 1960s (after Karen's death), and was subsequently presented to the Kenyan state as a gift commemorating decolonization. Colonial memories, lest we forget. This too won't pass and, fittingly, the building and its surrounding grounds were used to host a training centre for domestic workers – a school for servants. Afterwards came decay. Imperial debris in the midst of more colonial rubble.

The interesting thing is that the romanticization of this place, the construction of a very particular semiotic and emotional landscape, did not happen until the property was discovered as the antagonistic counterpart of the invented romantic home, the furnished and decorated movie site that enchanted everybody with its colonial props. With the help of a couple of thousand US dollars donated and invested by Universal Films, imperial debris was remade into its own simulacrum, a heritage site that

> features rooms designed in both the original decor and with props from the 1985 film. The grounds, which feature original equipment from the coffee farm, are also available for touring. The grounds of the museum are available to rent for weddings, corporate functions, and other events.[2]

What can be visited now may also be used as a site of still more heritage practices – family heritaging, remembering a wedding or a reunion, might be linked to a place where imagined characters engaged in imagined romance, surrounded by imagined people in an imagined Africa: There, everything seems blurred (Figure 4.1). Individual, personal heritage practices intertwine with practices of the colonial imagination. Blixen, as a

Figure 4.1 A vague tractor

destination for heritage tourism, here emerges out of a film set again and again, each time a blogger posts a series of images from the film, each time phantasies of love and safari trigger holiday plans, each time a travel guide offers a special suggestion. The actual building turns into a simulacrum itself, by being constructed as an emotional landscape that provides options for the heritage practices of diverse people and for diverse aims. Yet these largely seem to affirm heteronormative and neocolonial concepts of partnership, hospitality, entangled history, bodies and texts.

A Woman

Visitors appear to be less interested in a critical view of the colonial history of the place than in an affirmative performance of colonial melancholy. The visitors' commentaries in the museum's guestbook and the comments on webpages, as well as ways of consuming the site itself, reveal how closely personal experiences of the visit are connected to the stereotyped images presented in the film: of the colonial Other, of women, of bodies. Visitors take selfies wearing safari hats, sit on the manicured lawn on green plastic chairs arranged in circles and listen to a guide's explanations, or buy commemorative objects in the shop. Most seem to accept what is told them in the bits of text that are offered by the guides and on the small signboards mounted on benches and plants.

Notes in a field notebook:

This is the home of a woman whose books belong to Kenya's rich heritage, this deserves a visit, requires respect, elicits love.

The person whose spirit is still there (in the house) was wonderful.

She looked different.

She was different.

She talked to her Kikuyu servant as if they were coequal, coeval.

She called him when she needed to and he called her Memsahib.

The dresses are still there, too.

Costumes donated by the film company, or the director.

Doesn't matter.

But yet – to wear this: be this.

No critical question about colonial continuities, about what she had actually written in these books on Africa, or what others had written, Hemingway on the bedside cabinet.

Will read that tonight. – Noble prize winner. – Said she should have won. – Next time Steinbeck won. (Like him a lot.)

What did she write. Books on women, mostly.

A women's writer.

A writer who wrote about women, or for women, or on behalf of women? Very thin, eyes so big.

But the pilot. Chest hair coming out of a white shirt, unbuttoned.

Her writings sit there on a bookshelf. A large cabinet, rather. Filled with books. Translated into all the world's languages. Languages that count, at least. Not Gikuyu. Danish, German, Swahili maybe. The books must not be taken out; they are exhibited reliquiae. All she left behind. She never came back so that's all she left for them.

Them? Whom?

In the kitchen there is something we can buy. What was it? Coffee maybe, forgot what it was. Toilets behind labelled palm trees, sore feet. Sit where old men once sat who were still good enough for some work in spite of sore feet.

We have payed 1000 KSH in order to

> commemorate the life of Baroness Karen Blixen, the talented Danish author, poet and farmer. Karen Blixen is the author of several books including the famous Out of Africa later documented into a movie with the same title.[3]

Her servants, whom she treated well, called her *Memsahib*, and she called them by their names. Each of them had a name, a single name. They had single stories, too. Yet she wrote about African people as people who resemble each other. Children resemble each other when you watch them at play, siblings. Africans, she said, are free and close to nature and they will always remain that way and make their own decisions. They even decide when to die, like a little boy who died because he wanted to. This is a boy in one of her stories, *Kitosch*, who is beaten to death.

Ngugi Wa Thiong'o (1993) wrote about it and deemed this and other texts of hers unbearably racist, colonial, maternalistic. At this point in the visit, one is unbelievably grateful that he wrote this, as sad as it may be.

> Karen Blixen returned to live with her mother at the family home in Rungstedlund Denmark where she spent the rest of her days. It was at this stage in her life that she seriously started her literary career. Her first book 'Seven Gothic Tales[']' was published in America 1934, where it received the 'Book of the month' award. 'Out of Africa' was published in 1937. Karen continued to write a number of successful books and articles right up to the time of her death in 1962.[4]

An interesting perspective, too. All the experiences, emotions, romance, love, dangers, safaris, coffee plantation bankruptcies, open-air hair-washings and infidelities were achieved before she left, were incorporated while in Africa (Kenya), before she went back to Europe (the North). There, her writing took place. Raewyn Connell writes about the South as a site where data mining takes place in typical colonial and postcolonial settings, while theory-making and writing is done in the North. Africa, in these texts and in the museum, is presented as a source of something that is needed for the production of great colonial literature. It provides the raw materials – adventure, transgression. But the making of literature itself takes place elsewhere. And this is decisive; it is not so much simply the text itself as its production that reveals its ultimate coloniality.

Even though there is much in these texts that reveals irony and the subversion of hegemonic norms, a subtle critique of the British colonial society, there is never any unequivocal passage where colonial power inequalities are addressed as what they were and are – violent and ruinous. Importantly, the writer, the voice or narrator she creates, and her texts have previously been understood as hybrid, as being both European and African, male and female, North and South (Brantly, 2013). However, such hybridity emerges from performance, not the daily practice from which text and voice emerge. And this is precisely what becomes conceivable once the author's life is made visible as a simulacrum in the form of a Hollywood love story, which itself triggers the creation of yet another simulacrum, namely the remade house and museum. The ways in which the museum, the book and the film speak to each other, and visitors and guides get involved in this complex relationship, reveal how roles are already scripted, and how inescapable they seem. Suddenly, Blixen's insinuated hybridity and interest in the colonized is revealed as colonial mimicry. The mimicry is such that the ruinous and violent colonial actor turns into the colonized as a hero and deified figure (Roque, 2020), yet performing this very different role only for some time, not more. There, at the fringes of the city and of everything, where lions could (then) be seen, one could be somebody else, could live a different life.

> Looking back on her life in Africa, Karen Blixen felt 'that it might altogether be described as the existence of a person who had come from a rushed and noisy world into a still country'.[5]

'Karen is a way of life', Danielle de Lame (2010) notes in *Grey Nairobi*, and the museum and its grounds are deceptive in suggesting that there might have been any real hybridity. Not hybridity emerging from a white body immersing into what Blixen calls the essence of an entire continent, but something more substantial, less exoticist, less Orientalist. But Karen, de Lame writes, remains remote and exquisite, like the segregated flowers in the museum garden.

Hybridity is, we assume, something that has to do with the North rather the South. This might astound some: certainly, the colonized world has always been seen as hybrid; the colonial towns of the tropics are hybrid (and vibrant, don't forget vibrant; warm and moist, too – this is how hybridity is made to grow best), Creole languages are hybrid, postcolonial societies are hybrid, and so on. But these are hybridities that are constructed as being salient (and therefore in need of explanation) by means of positioning them in opposition to an imagined orderly (non-hybrid) North. The nation state is orderly, as are democracy, Northern road systems and classrooms – they are not hybrid.

The Southern (African, and therefore Kenyan) players that feature in Blixen's text, the film and the museum never offer any mooring for hybridity. They are fundamentally defined, fixed and named. The servant Kamande Gitura is forever seated on that stone bench, is forever just Kamande Gitura (even though he might well have had other names, simply by being a whole social being), a cook, a Kikuyu. Nothing mixes or mingles here. A man like him would not have been portrayed as a partner of a woman from Europe, for example, and he alone, without any link to the writer, would perhaps not have been interesting enough as a figure in a film, text or museum that 'commemorate[s] the life of Baroness Karen Blixen, the talented Danish author, poet and farmer', would never have been more than just a faithful piece of decoration. He is part of the staffing needed in heritaging. Did he have a pseudonym? Were there nicknames? Not unless they were chosen by the European chronicler, we assume.

But Blixen had names, and used pseudonyms – abundantly so. She is hybrid, and she claims authorship over her hybridity, challenging Northern bourgeois order. This is not inflicted upon her, by colonial dislocation or by disruption, but chosen and performed. And we simpletons wonder a bit: is this the same Blixen as Tania, who wrote this book, which inspired the film? Yes. And of course we always thought that. Easy to guess, actually. But then: 'I find it difficult, however, to accustom myself to Karen Blixen's many names, e.g. Isak Dinesen and Pierre Andrezél besides the American synonym Meryl Streep', Søren Rasmussen writes, in *The Greatest Safari*. And the website of the Karen Blixen Museum in Denmark offers a dozen or so more names to add. This is hybrid, this is enigmatic. Self-authorship in order to dissolve. A photograph of her taken by Beaton bears her signature on its back: 'With love, Tania'. This is her German *nom de guerre* and she writes it on her very British portrait. She is said to have been subversive, against British upper-class etiquette.

At the museum, we are offered photographs of her. Few of them show much of her connection with Kenya (Africa). Images that illustrate her career as a writer abound. She is shown with many celebrities of her time, and all we see on these images are White people. White people who write and read about Africa, as we do, as tourists do, as one just does.

But what captures our eyes more than the Whiteness of the place and its imagery is its dryness. This is a very dry place. Dry things to see, dryness in what we hear. Also the dried-out body in these photographs: this woman is so thin. Eyes too large. We look for explanations: after syphilis, she depended on laxatives, some of her biographers observe. Became addicted to laxatives. Couldn't eat; probably anorexia. Died of malnutrition. Beaton's portrait is hard to bear. Did he care?

Automummification is an extreme form of asceticism. In religious contexts it was, or maybe is, not considered as an act of suicide, but as reaching a form of enlightenment.

And mummified language?

The absence of embodied presences is something we find distressing. All these stories about love and relationships, and then not a single trace of language to make this happen. In the film, the other woman, whose lover dies as well, the other BLACK woman, remains mute. In Karen's house we are reminded of her piercing eyes, her quiet stare. Here, too, nothing speaks. Linguistic drought. The volumes of *Out of Africa* are on display, in all the different languages into which it was translated, framed book covers, written explanations on objects, even on pots and boxes. Everything is labelled and defined, in English, sometimes Danish, everywhere. But no interaction takes place: this is language that does not reach out. It does not speak.

What we can hear is the pump behind the toilet, after flushing. Blixen was addicted to laxatives. The tour guides working today explained what needs to be explained outside, speaking with soft voices, sitting on green plastic chairs that stand on the immaculate lawn. Inside, it is tidy, orderly, quiet. How do we revive this dead matter?

What remains: language conserved, inarticulate and remote. A Hollywood movie, its image politics and Eurocentric storyline. Signboards of Pentecostal churches on the road we took coming here. What is the reason for visits to such places?

Notes

(1) See https://hookedonhouses.net/2010/03/22/karen-blixens-house-in-out-of-africa/ (accessed 12 August 2020).
(2) See https://www.kenyamuseumsociety.org/karen-blixen-museum/ (accessed 12 August 2020).
(3) See https://www.kenyamuseumsociety.org/karen-blixen-museum/ (accessed 12 August 2020).
(4) See https://www.kenyamuseumsociety.org/karen-blixen-museum/ (accessed 12 August 2020).
(5) See https://www.kenyamuseumsociety.org/karen-blixen-museum/ (accessed 12 August 2020).

5 Highway to Hell

Transition Grammar

In a tourist bus with dark windows that goes from here to there, and there to here, people sit and watch what is passing them by outside. Outside, there are other people, and they do not look back. There is no use in looking back; one such bus after another, making some noise and leaving some dust behind, like the wind that blows into the plastic bag heaps from time to time, all day. There is work to do, a road to travel, business to mind.

Not that those in the bus do not also have work to do, a road to travel and business to mind. But in order to do the things they have to do now, they need those passing people outside and their plastic bag heaps and dust and stillness. While the aircon hums, the click of a camera resounds. Click-click-click, and soft voices speak. Faces turned against the darkened glass, eyes pressed against the windows: 'Look here!' A line from a poem by Anne Carson (2016) reads "I suppose and I presume."

It belongs to a poem named 'Sonnet of Exemplary Sentences from the Chapter Pertaining to the Nature of Pronouns in Émile Benveniste's *Problems in General Linguistics* (Paris, 1966)', which can be found on the page opposite another poem that is called 'Sonnet of the Pronoun Event'.

In the tourist bus with its dark windows and the closeness of faces and glass, the work that people do is that of supposing and presuming. Also the talking: this is another work that should be done on the way to the destination to which they are heading. Pronouns indeed play a role in it, and an eventful role, even. To the linguist, the language of the interior of the tourist bus on the way to its destination is an important field of study, which helps to understand the use and meaning of pronouns in particular, as well as other communicative phenomena such as click sounds.

In the air-conditioned tourist bus with its dark windows, there is some seclusion that shelters those who speak from those who might want to ask. The driver may not even hear what they say, with the aircon humming all the time. Like water in a park's refreshing spring, it hides the meanings of the words that are spoken.

In this seclusion, which turns the interior of the tourist bus into a kind of liminal space through which the tourists will have to pass until their destination is reached, a special form of language is used, as is normally the case in liminal spaces. In the tourist bus, this is a language that is used to suppose and presume. Therefore, its vocabulary is a bit reduced; it is a specialized language that fulfils very particular tasks, that is not used to bargain in a market, for example. Verbs in this language form a closed word class, which is typical of some ritual languages, whereas in languages of everyday communication, verbs often form an open word class (Kanuri is among the exceptions here; cf. Dimmendaal, 2006). The verbs that form part of the tourist bus interior register include:

(1.a) schau! 'look.IMP'
 denk! 'think.IMP'
 überleg! 'consider.IMP'
 sieh! 'see.IMP'

(1.b) leben 'live'
 wohnen 'inhabit'
 zurechtkommen 'get along with'
 gewohnt sein 'be used to'

In Example (1.a), a group of verbs denoting cognition is presented. An interesting feature of the verbs in this category is that they are often used in a truncated form, almost in an iconic fashion. The shortness of the verb parallels the limitations of cognitive experiences in the liminal space inside the tourist bus. Mostly watching what is outside, the people inside experience visual sensations and imagination. Nothing else.

In Example (1.b), some verbs that express stative location and habituation are presented. The verbs of this particular set are hardly ever truncated, which might be another example of iconicity in ritual language: the everlasting state in which those outside seem to be requires long verb forms that provide an almost tangible sensation of continuity.

An interesting feature of the verbs in group (1.b) is that they tend to occur with only one subject pronoun, namely *sie/die* 'they'. In other words, almost all constructions that pertain to life and existence are constructed in the third-person plural. The outside world viewed from the tourist bus is not perceived as fragmented and diverse, and all contradictions that are part of a life shared by many individuals are erased from discourse. The people who can be seen outside are constructed as one large group, a mass, a single collective body.

Unsurprisingly, collective nouns represent another word class that is characteristic of the register of the tourist bus interior:

(2) Hiesige   'autochthonous people'
    Afrikaner   'Africans'
    Locals   'locals'
    Leute   'people'
    Eingeborene 'natives'

Anne Carson's 'Sonnet of the Pronoun Event' ends with by asking whether pronouns are like lilies; she concludes that they indeed are, how else would they constitute events.

Events can happen in unexpected ways, very much to the surprise of their beholders. In some languages, actually many of them, this is encoded by the use of specialized grammatical forms (Aikhenvald, 2004). Often described as a sub-category of evidentials (which largely encode the source of information), miratives express unpredictability and surprise. That miratives and evidentials are often in close relationships with each other is not surprising at all, because evidentials also help to describe the reliability of information, as first-hand information or as something one only knows through hearsay. Miratives express that something has really happened (truth, a particular quality of information) even though it seemed improbable and lily-like.

In the language of the tourist bus interior, miratives abound. They do not form a special class of grammatical formatives, however, but are mostly derived from interjections and (much more rarely) adjectives. As one can expect, they are used to modify verbs that express vision and cognitive processes based on visual sensation. Examples are:

(3) boah   kuck   'surprise: look.IMP'
    MIR.VIS   look.IMP

    nee   ne   schauma   'unbelievable: see there.IMP'
    INTENS   MIR.VIS   see.IMP.DIR

    hammer   glaubstes   'impossible: do you believe this'
    MIR.VIS   believe.2SG.O3SG

The use of mirative constructions is constrained by gender. Female speakers tend to utter them more often and in a more nuanced way, whereas male speakers are reluctant to use miratives extensively. If they do, they usually utter them in a hissing way, with more friction:

(4) Gendered miratives
(4.1) Female mirative register
heyyyy sieh doch nee ey 'improbable: just see.IMP'
MIR.VIS look.IMP INTENS INTENS INTENS

(4.2) Male mirative register
ssss nee hassse nich gesehn 'improbable: just see'
MIR.VIS INTENS have.2SG NEG see.PAST

In discourse, mirative constructions tend to introduce comparative constructions. Moreover, a characteristic feature of the register of the air-conditioned tourist bus is that comparative constructions often merge with locative constructions. We interpret this as another highly iconic function of the register: spoken in a fundamentally transitional, even moving entity, comparisons between different places – here and there, then and now – are of utmost importance. Examples of how the dynamic state of those who speak is represented by language are:

(5) Locative comparative constructions
nee ne glaubstes wie da in Italien der Basar
INTENS MIR.VIS believe.2SG.O3SG COMP LOC LOC Italy DEF market
'Hard to believe that this is like there at the market in Italy!'

boah kuck wie in Pattaya
MIR.VIS look.IMP COMP LOC Pattaya
'Just look, the same as in Pattaya!'

While toponyms and other proper nouns form an unstable part of the tourist bus register, loan words and neologisms that refer to foreign places and things do not. Dynamicity and transition are embodied and performed in the bus, and this is saliently reflected in language. Words that have origins in other languages are used extensively. They are somewhat isolated as entities, as items worshipped in a cargo cult, and usually do not occur together with other lexical entities of the same origin. In the tourist bus language, solitary words help to stylize and perform otherness and transition. Some examples are:

(6) Dashiki   'cotton shirt with African print'
    Kikoi   'cotton cloth with African print (*kanga*)'
    hakuna ma Vodka   'cheers' (< *hakuna matata*)
    Zungu   'white person' (< *mzungu*)
    Pumpe   'beer' (< *pombe*)

Unlike the other forms presented above, these examples are characteristic of reverse transition, i.e. of travel back to the airport from the resort. They might serve apotropaic purposes, such as shouting at the foreign and Other in order to chase them away. Once the speakers leave the bus and enter into another form of transit – airport transit – they change their speech register again, often interrupted by bouts of silence.

In the poem 'Sonnet of Exemplary Sentences from the Chapter Pertaining to the Nature of Pronouns in Émile Benveniste's *Problems in General Linguistics* (Paris, 1966)', Anne Carson mentions eating a particular fish not necessarily involves knowing its name.

Roadside Narratives

18 April 2018 – Likoni – Tiwi – Ukunda: Three locations in Kenya. These toponyms are normally found on matatus, the small buses that run between these places, and serve as directional markers. They demarcate places along a road in Kenya that doesn't naturally meander, but rather follows the coastline rectilinearly from Mombasa southwards, having probably destroyed natural habitats during its construction. Rectilinearity is a shocking and destructive term in regard to nature and humans. It exposes the vulnerability of togetherness by presupposing that if someone or something is in the direct way between two places, it will be cut apart. For this particular road, this scenario probably took place when the road was constructed, but since then, people have integrated the road into their daily lives and their neighbourhoods. The land on both sides of the road belongs to the government, and people living in the area are supposed to know the width of the road reserve. *The Standard*, a Kenyan newspaper, writes in 2010 that: 'the common road sizes (road carriageway plus road reserve) are nine meters, 12m, 15m, 18m, 25m and 30m depending on the opinion of the physical planner and local authority's town planning department.'[1] The varying distances illustrate the discretion of the government and the understandable lack of knowledge of the appropriate measures by the people living along the road. People have thus started to build houses or shops along the road, ignoring blueprints and rules concerning government land. What is a livelihood for many people is a thorn in the side of officials. For decades, people have been constructing their stalls along the road, depending on finding a good place in order to earn money.

On 18 April 2018, excavators and bulldozers dug a swath of destruction approximately 10 km along the highway on both sides. The owners had been informed three days in advance that they had to destroy their property themselves; otherwise they would be charged for the demolition. Therefore, the owners of a private school who had set up their sign where many other signs had been for years went and demolished it (Figure 5.1), while the neighbours watched them, hoping for the best and not thinking about demolishing their own livelihoods (Figures 5.2–5.7).

Highway to Hell 61

Figure 5.1 Owners of a school destroying their own road sign

Figure 5.2 Remnants of a house partly built on governmental land

62 The Impact of Tourism in East Africa: A Ruinous System

Figure 5.3 Remnants of another house partly built on governmental land

Figure 5.4 Demolished by bulldozers: the wall of a school in Tiwi

Figure 5.5 No houses left

Figure 5.6 Demolition of homes

64 The Impact of Tourism in East Africa: A Ruinous System

Figure 5.7 What is left: coloured walls of former happy homes

The Star describes the day of demolition as complete chaos:

> Traders at Kongowea market incurred losses running into millions of shillings on Thursday after the highways authority demolished their stalls. Some of the business people at the market's Lights section broke into tears after their pleas to the Kenya National Highways Authority fell on deaf ears. This section of the market accommodates around 1,500 traders. The victims got to work early as 5 am hoping to earn their families some money. Hell broke loose at around 7.30 am when a bulldozer's engine roared to life, flattening everything in front of it.[2]

The legislature clearly sits at the longer end of the lever and the ruination of families will soon be forgotten.

What follows are some photographs that record the ruination along a touristic road, the road from Mombasa to Ukunda, where thousands of people spend their holidays every year, passing along the road exactly four times during their stay: on their way from and to the airport and to and from a National Park (with the exception of those who visit the Masai Mara National Park, as they normally take a flight from Ukunda). Tourists are not informed about the topic of the road reserve demolition and watch the scenery without much emotion, albeit with astonishment (see Examples 3–4 above).

The roads are clean, clean and rectilinear, cutting through the private lives of individuals. One can see the inside of privacy, the painted walls as they shine in orange and purple, like posters without a message. The openly depicted intestines of privacy let curious others reflect about what has been there before, who has lived in there, who decided to paint a wall yellow. Perspectives vary in this scenario of ruination. Small-scale business along the road is destroyed, giving way to unproblematic transport, both national and international. We asked a man who has a security company in Diani, not far from this area, about the reason for the road demolition. He writes:

> Regarding the road issue, it's been said many times that the roads infrastructure needs to be improved so that trade with our regional partners, within Africa who are landlocked can have their cargo delivered without much challenge as it is at the moment. The other reason is to improve on the flow of traffic as there are many cars in the country but the roads are still not big enough to accommodate all cars at the same time hence the usual traffic jams, so the only way was to have the roads expanded by using the road reserve and anyone on the reserve was removed by force as it's government land.
>
> The demolitions were very necessary in the first place even without the expansion as the road reserve which is government land meant for road expansion had been given out illegally by corrupt officials who never thought that the government would catch up with them at one time. The other people who had their properties demolished were the ones who interfered with the riparian land. This is where our natural rivers used to flow but again the greedy and corrupt officials again sold this land that is close to the rivers and affected their motion or movement. The government has been demolishing most of these buildings and unfortunately the prestigious shopping malls have been caught up in this demolitions as some of them were on the riparian land.

Since the pictures were taken, not much has changed. People have collected their corral stones and stored them properly in order to be safe from theft. Nobody dares to start constructing again. Salespeople have taken up positions to sell their goods without a stall. A year and a half later, broken walls stand amidst lovely vegetation which slowly repossesses land. Other demolished buildings are freshly painted in turquoise, yellow and pink, as life goes on. There is no sign of road work, anywhere, which once was the reason why people were made homeless.

Notes

(1) See https://www.standardmedia.co.ke/business/article/2000023256/establishing-your-plot-boundaries (accessed 12 August 2020).
(2) See https://www.the-star.co.ke/news/2018/02/22/kongowea-traders-break-into-tears-as-kenha-flattens-stalls_c1719106 (accessed 12 August 2020).

TOURINATION CHAPTER SIX

SUPIDUPI

RUINS ON THE BEACH

RUBBLE ★ DUST ★ DEBRIS ★ CORPSES ★ SHATTERED DREAMS ★ FAECES ★ TORN PAPERS ★ GARBAGE ★★★

6 Ruins on the Beach

Old Sites

Images of Kenya's coastal regions are usually images of idyllic beaches. Endless stretches of white coral sand, turquoise blue water and shady coconut trees. No people. These images always seem to accompany promises of exclusiveness and access to otherwise closed territories – premium lounges, upgraded suites, sea view cottages. They emerge from ads and portable media, smartphones on the train during a tedious ride through grey suburbs. Language turned into slogans, spaces to fill an online form, flights quickly booked. A trip to Kenya, deserved and achieved.

While a few premium resorts around Diani affirm phantasies of deserted beaches as spaces of prestige and elitism, most other hotels and beach clubs do not. Already before the Covid-19 pandemic, they have been closed down in the course of a giant crisis in tourism, economy and politics, and have begun to fall into ruins. Outside the premium littoral, this creates fears of danger. It is not only the guidebooks that provide orientation in this space that no longer seems to offer any surprises, but also the many signboards that are posted everywhere, that ask us to beware – of the young men, of falling coconuts and rubble, of polluted water and venereal disease.

At the beach, ruins look astonishingly similar. As if the architecture of the non-place was so uniform and so reduced to just one effect – providing space and opportunities to tourists for their consumption – that the decaying matter left behind had no chance to turn into anything at all. Just rectangular shapes and circular shapes and blocked entries. Blocked exits. Empty pools that get filled again: with rain, rodents that have fallen into the rain water and insects that feed on the dead rodents. Broken windows with mummified frangipani blossoms and dusty cobwebs. Musty mattresses on filthy bedsteads. The only traces of human life spent and lived seem to be faeces that reek in the sun.

Some of the ruins exhibit ruined animals (Figures 6.1–6.6). Giraffes that peel from the walls of an old casino, an elephant on a door, a buffalo on a car park. Hollow eyes, greyish coats, missing legs. Ears gone by. These animals have always been dead, and it is only now, surprisingly, as everything falls into ruins, that they come alive. Like their cousins in the

Figure 6.1 A vague giraffe

Figure 6.2 Giraffe and tree

Figure 6.3 Elephant, trees and palms

Figure 6.4 Elephant and pool (dry)

Figure 6.5 Buffalo and rhino amidst scenery

bush, they are shades and spectres that can be glimpsed at dusk. Vague sensations of other life that is also there, besides us. And as plants and cobwebs and other forms of life take over and transform the hellish nonplace of this tourism paradise into ruined space, the animals slowly come to life, are inhabited by other animals like birds and centipedes.

Figure 6.6 Numbered elephant

A long time before this all happened, and before the roads were made wider and better and tourist complexes were built all along the coast, on littoral space, the area was covered with forest. What remains, as protected areas under UNESCO patronage, are sacred sites such as Kaya Kinondo, sacred forests. Everywhere else along the road leading south from Diani to the Tanzanian border are hotels, resorts, ruins of hotels and resorts, houses, ruins of houses, farmland and ruins of farmland. Villages remain 'authentic' and 'rural' because the only industry that money was put into was the tourism industry. Young men who have grown up in these villages hang out amidst all the ruined space at the beach and look after small businesses, or simply try to sell themselves, as there is nothing else there that can be sold.

Turning the gaze to ruins and ruination allows us to grasp the durabilities of marginal placement of particular people, namely people who, lacking access to particular economic resources, are considered 'local'. The construction of local young men as jobless, failing and criminal can also be seen as the continuation of inequality as a consequence of previous injustice and violence, of various kinds. The ruinations of places and people are ongoing processes that elegantly have been framed by the anthropologist and historian Ann Laura Stoler as the durabilities that imperial formations produce. Stoler points out that there is a 'relationship between colonial past and colonial presents, [and the] residues that abide and are revitalized' (Stoler, 2013: 5). The products of ruination are therefore complex and diverse, and while the residues of empire have often been assumed to consist of utterly tangible remains, Stoler argues that there are also other ruins: 'If the insistence is on a set of brutal finite acts in the distant slave-trading past, the process of decay is ongoing, acts of the past blacken the senses, their effects without clear termination. These crimes have been named and indicted across the globe, but the eating away of less visible elements of soil and soul more often has not' (Stoler, 2013: 1). Imperial debris therefore consists not simply of neocolonial hotel buildings rotting away in the tropics, or of the creation of world heritage sites that are what remains of a more complete social life, but also of ruined lives, of 'persons who become "a human ruin", "leftover" in their unexceptional, patterned subjection' (Stoler, 2013: 23) to abandonment and exclusion. In other words, ruination, the production of imperial debris, can result in poisonous environments, such as the coral reefs and marine ecosystems off the beach that have been destroyed by sand dredging, exploitation and pollution, but also in neocolonial economies that exclude people from sustainable incomes, and in images of young people as risk factors to social balance and order. 'To ruin', Stoler says, needs to be framed as a very agentive verb, an action that is powerful and that can be controlled by particular agents. Its duration, however, is out of the control of individual actors, she suggests: race is imperial debris, as are resentment, dispossession, and violent environments created through the uneven reallocation of access to resources.

The young men who hang out in the ruins that were once profitable all-inclusive resorts ask for a bit of attention: a greeting, a bit of a shared walk, time for a chat. Human ruins, maybe, depending on how the subjective definition may allow us to judge the life of the local tourism workers. The term is not applied to the characteristics of the people, but rather to the life they had to adjust to after the income from tourism drastically decreased after its breakdown.

The topic of ruins, strangely, calls for images rather than words. Images of what this might look like – castles over the sea, expectations that hail from encounters with romanticism and gothic novels. Even industrial ruins and other recently lost places appear to produce mainly images, not text. The bizarre aesthetics of transformation ask for very particular genres – the coffee table book, the exhibition catalogue, the blog. And while images elsewhere might have the power to reveal, here they conceal and leave us daydreaming of colonial expeditions, discoveries of hidden treasures, precious lost things underneath the rubble.

It is striking that disadvantaged 'ruined' people are missing in the Kenyan world of high-class romantic daydreams, although they play a vivid role and are definitely more noticed by tourists than the stone ruins. The visual representations of the Kenyan poor are reserved for the oversized posters of NGOs that fill European cities at Christmas time. There, they form part of the colonial mechanisms that endure as they help us to construct, belittle and advertise the rest of the world. It remains a matter of space to transport what is to be transported.

The images of ruined space that are produced betray its temporary form and the transitions that take place. These ruins are fixed and stable, on high-quality paper and in nicely edited formats. But how can we speak or write of them? What kind of story emerges from the ruin?

> While the older conceptions of rootedness and autochthony seem intellectually bankrupt, the heady theories of creative metissage have run aground upon the rocks of contemporary reality. (Greenblatt, 2010: 1)

This is what Stephen Greenblatt (2010) has to say about the urgency to 'rethink fundamental assumptions about the fate of culture in an age of global mobility' (Greenblatt, 2010: 1–2). In *Cultural Mobility*, he reminds us of the 'restless process through which texts, images, artefacts, and ideas are moved, disguised, translated, transformed, adapted, and reimagined in the ceaseless, resourceful work of culture' (Greenblatt, 2010: 4), which 'obviously long preceded the internet [...] or the spread of English on the wings of international capitalism. [...] The apparent fixity and stability of cultures is, in Montaigne's words, "nothing but a more languid motion"' (Greenblatt, 2010: 5).

What about language? At the remaining resorts, frozen in languid motion, guests lounge around pool areas, in restaurants and on balconies and stare at smartphones. They type a few words that form part of an

Instagram story. Shoo away the waiter. More consumption: messages on the weather and the 'locals', on beaches and barbecues and home.

The smartphone has been seen as an agentive instrument to speed up both ruination and motion, through an incredibly fast distribution of images, ideas and language. Its materiality transcends class boundaries, as the tourists in the resort type their time away, as do those who wait for them outside the resort's boundaries, in taxis and *tuktuks*, at the beach and on the road. As a portable portal to the internet, the smartphone is increasingly thought to enable marginalized people to play central roles in contributing to social good across the globe and in enhancing change, as in the case of the activist and Iranian refugee Behrouz Boochani, who received the prestigious Victorian Premier's Literary Award in 2019 for his novel *No Friend but the Mountains*, which he wrote during his incarceration on Manus (where he remained for more than six years) with the help of a smartphone, as a series of seemingly endless WhatsApp messages.[1] The smartphone obviously does more than merely linking migrants and places, enabling underprivileged people to obtain information, create networks and access foreign language (e.g. Beck & Beck-Gernsheim, 2014; Bloch & Donà, 2018; Deumert, 2014; Leurs, 2017; Lourente, 2017); it turns into a powerful instrument for changing the precise context out of which its necessity emerges, as a tool for southern critique on northern imperialism (Boochani, 2018: 20).

Yet this picture bears a strange similarity to earlier images of the colonial Other as an agent of humanism. It is not surprising that many of the published statements lauding Boochani's work highlight the role of his translator, to whom he sends his texts and who translates them from Kurdish into English and from WhatsApp into literature. But what happens to cultural products that travel through time and space to emerge and to be enshrined in new contexts and configurations?

Old Roads

Elsewhere, on the Mediterranean island of Mallorca, ruination has turned into performance and the smartphone into an amplifier. The beaches close to the island's airport are popular with party tourists, who mostly come from Germany and Britain, often only for a couple of days, in order to engage in the transgressive amusements of carnivalesque spaces. For some time now, the streets in front of the clubs and the beaches nearby have also attracted migrants from West Africa, who mostly come from Senegal and Nigeria, often for a couple of years, in order to participate in the petty economies surrounding the party space. As street vendors of cheap souvenirs, they are integrated by their customers into a performance of inversion and possession. German party tourists, who often wear colourful costumes such as party hats and motto T-shirts and mimetically perform clichéd lower-class practices, claim the island as a German colony, audibly in party songs and visually through text on bodies and buildings. The West African street vendors, similarly dressed in a carnivalesque manner, turn into a southern alter ego in this play, and into theatre props. They are uniformly addressed as *Helmut*, which is a common although old-fashioned name in Germany, and their images are circulated in this stereotyped representation through various Instagram and Facebook outlets (Nassenstein & Storch, in press). Mocking the Other's Otherness is a play, everybody claims, which is, of course, organized according to the desires of the paying customers. This is without doubt obscene and transgressive, a performance of racist stereotypes and colonially based power. And in addition, the drunken tourists celebrate themselves with photos of their drunken companions.

Yet most of the Senegalese and Nigerian migrants prefer to convey different images, as the Nigerian-Spanish activist and entrepreneur Festus Badaseraye observes: 'Europe is paradise to most Africans. Europe has a better system than Africa. We don't have a social security system, look at our schools, roads, hospitals. Everything is better here in Europe' (personal communication, March 2019). The experiences of exclusion, abuse, exploitation and lack of success are hardly ever communicated to others, and likewise most migrants keep silent about bouts of depression and angst (Lindtner, 2018; Smith, 2019). There is a sense of shame, M., a man in his mid-thirties said, in being forced to live in such a situation, while people back home, who have invested so much, expect success stories. A., who has lived as an undocumented migrant almost exclusively in the party zone, where he feels safe from the police, says that he prefers to send home images of clean, quiet places on WhatsApp, the opposite of the noisy working contexts in which he usually operates.

Among the migrants themselves, however, different material is circulated. While waiting for customers, one cannot help watching what happens: the decadence of intoxicated tourists who seem to have too much to spend, heterosexual White males in a heteronormative White male space, where women are gang-banged in the cheap triple rooms of the Hotel Niagara, or where one can afford a cheap blow job by one of the women working in the streets nearby, where migrants are integrated into the party space as complicit persons who can be molested and abused in a drunken performance of superiority. And one can make short films of what one sees, because the superiority is such that the superior appears not to be afraid of the defacement. The little videos that are circulated, among those who might have come there expecting other sights than these, are videos about temporary but crucial losses of control.

The first video was of a young man in swimming trunks who jumped from a hotel balcony into the pool, narrowly missed it and broke his hips on the concrete. *Balconing* has claimed several deaths among intoxicated tourists in the past few years. An anti-tourism graffiti in Barcelona reads *balconing is fun*.

The second video we see shows 12 seconds in the life of an obese young man who wears an orange party wig and matching Borat swimsuit, sunglasses, purse and black sneakers. He walks down an average road (not a party boulevard). One hears men who are softly laughing.

Then a film made at night. A young man has vomited in front of a grocery shop, just in front of some waiting taxis. He went into the shop, bought a toothbrush and toothpaste and cleaned his teeth, standing on the sidewalk. Other men try discreetly to push him aside, away from the shop. He moves only a little, then throws toothbrush and paste on the street, spits and moves on. Laughter, expressions of surprise and disbelief.

The next video shows a street at night and lasts 31 seconds. A man stands on a sidewalk and masturbates into the mouth of a woman who kneels in front of him. A voice repeats, in disbelief, *Esto lo que pasa en El Arenal. Que grande*. Most of the spoken comments on these videos are in Spanish (never Mallorquin), which is used in leisurely situations among many long-term migrants from West Africa. And then, 'very sorry [...] this happened in Arenal' is written in the text message underneath. In the party zone, there are signs indicating high fines for public sex.

And so on.

Nobody understands this drunkenness, we are told. Yet these films are comparably peculiar, particular and local to what is circulated by the tourists, whose posts we see as mimetic practices, playing with Othering. We suppose that these West African videos are more complex than this. In an environment where one is largely negated as a coeval person, is denied the rights of hospitality and citizenship, discourse is more fundamentally annihilated. Conversation is replaced by declaration, addressing by naming. Mimesis runs aground.

A., a Nigerian lavatory attendant in one of the clubs, reports that many of her clients are excessively drunk when they use the toilet and often need to be washed and helped. She supplies the women with deodorant sprayed under the arms and between their legs and with anything from chewing gum to make-up to 'civilize them again', as she says. She regains control over her working environment and earns some tips as added value. She and her colleagues find the tourists lacking in education, an opinion shared by many of the street vendors outside. The migrants in the party space claim the task of providing this, as Behrouz Boochani claims that he and other refugees return humanity to Australia.

The images, therefore, are the symbolic proof of a denial of hospitality that creates failed hosts, turning them into obscene people and pitiful sights. Obscene visual material here obtains the status of a magical object, turning the artificiality of the smartphone and the signs it produces into a powerful instrument that is able to make meaning of one's place in the world. The philosopher Souleymane Bachir Diagne has made this a bit more transparent in his (2013) dialogue on truth and untruth. Bachir Diagne is interested in the cultural mobility of ideas and in how pre-existing or local thought is integrated into Islamic religious practice in Wolof society. The deeper meaning in the folktales about experiences of the appropriation of the local into the global, he suggests, is that they tell us that 'belief leads to truth, although what we usually think is that truth leads to belief' (Bachir Diagne, 2013: 5). Inhospitality and accusations of sorcery – practices directed against the abject Other – are practices that oppose the norms and values of religion and relate to local practices and local systems of belief, he suggests. In this constellation,

> truth is the path that exists, the untruth is what deviates from the path which is delineated. [...] So the truth is not only what is. It is also what should be. Consequently, when someone does something wrong, he is told that what he did is not the truth, it deviates from the path. (Bachir Diagne, 2013: 11)

In a non-discursive setting, where there is no place for such a conversation, the digitally circulated obscenity turns into magic that serves as a protection from precarity and debasement. And because religious truth does not nullify the local – tradition, as Bachir Diagne writes – it needs to replace it as well as to renew it. Civilization workers are now able to do their task:

> Former President Senghor often used to say that the people's tradition is not something that [religion] should displace: it should melt into it, penetrate its essence. [...] Then we can come to an agreement that tradition is not enough for us to talk about truth, we should say making sense and agreeing with the times. (Bachir Diagne, 2013: 13)

Whatever is undesirable that the local, the traditional, holds in store, there is the endless potential of culture – religion – to transform itself and that with which it is confronted. The abyss of the internet, the ubiquity of social media, are not given much meaning here. What counts is transformation, it seems, which is based on time. 'Not language, only process', a Nigerian woman, who talked about her migration into all this, told us.

We end this with two suggestions of how to come to a deeper understanding of the transformative power of that which is shown and that which is not shown. We end with two different thinkers, a Senegalese philosopher and a British historian, who come to the same conclusion:

> Mobility studies [...] are essentially about what medieval theologians called *contingentia*, the sense that the world as we know it is not necessary: the point is not only that the world will pass away, but also that it could all have been otherwise. This *contingentia* is precisely the opposite of the theory of divinely or historically ordained destiny that drove the imperial and figural models of mobility. [...] How is it possible to convey a sense of *contingentia* (and its counterbalancing illusion of fixity) in practice? [...] Montaigne's response was not to try to construct an abstract system but to try to describe a single object in motion, himself. (Greenblatt, 2010: 16f.)

> I will stop our conversation here then. There may be some other thoughts that occurred to me during the conversation, such as the meaning of 'time' ... We will just have to leave it at that for the time being, as it is getting late. [...] [L]et us go our way. (Bachir Diagne, 2013: 13)

78 The Impact of Tourism in East Africa: A Ruinous System

From here on, there remain images of what can be shown of a walk from the beach back to the airport, from where tourists return home and migrants are deported to what might not be a home any longer (Figures 6.7–6.12).

Figure 6.7 Ad and palm tree

Figure 6.8 Road and pine trees

Figure 6.9 More pine trees

Figure 6.10 Pine trees and cactus

Figure 6.11 Windmill ruins

Figure 6.12 Highway bridge

Note

(1) See https://www.bbc.com/news/world-australia-47072023 (accessed 12 August 2020).

CHAPTER 7

Relocation and Relationships

Explore the explored
Know the known
Name the named
Eat the cooked
Spurn the raw

7 Relocation and Relationships

Encounters on the Way

Years back, during childhood days, the radio in the car would blare strange words into the hot dusty summer afternoon. *Heia Safari*, a male voice crooned, and what that meant, and was, never became very clear to the child on the back seat of the car. The father, a lover of jazz and Bebop, hastily switched to another station, obviously finding *Heia Safari* unbearable. Sometime later, a find in the great-aunt's attic: *Heia Safari*. A book on Africa, written for the youth.

Years and years later, a man waited in front of the Kenyan beach resort and said: *Safari ham wir schon?* 'Safari done already?' The hotel's souvenir shop offered various items that bore images of the 'big five', the large mammals that are presented as the highlights of all the sightseeing and photography. A decorative plate with lion, buffalo, elephant, rhino and leopard would later testify that nature had been seen, a journey had been made, wilderness conquered.

The man continued his attempt to establish a conversation by asking a few more things about how we liked it here and why we had skipped the national park. Most tourists came on a package tour which usually included a safari in one of the parks first, with animal sightings and Maasai village visits, and then continued with a week on the beach, with an all-you-can-eat buffet and pool gymnastics. Not easy to get into contact with the tourists, he explained politely, who do not seem to trust the beach vendors and local entrepreneurs and who do everything with their tour guides. He was from the village that was no longer a village because it had been replaced by beach resorts and tourist malls. The rejections of the majority of the package tourists felt like a second expulsion from home, after the expulsion from the land itself. It created frustration and violence in the ways in which conversations at the beach were ended, indirect communications in German and English, but rarely Swahili. The things that were felt and perceived remained unsaid, as if these homeless conversations had nothing to do any longer with the feeling of a place, the sensual perception of the bodies that populated it. Yet the language of the senses is ubiquitous in the safari world, which remains a colonial sphere.

Sensual Stranger

'My young head was in a ferment of new things seen and realised at every turn and – that I might in any way keep abreast of the flood – it was necessary that every word should tell, carry, weigh, taste and, if need were, smell' (Kipling, 1990: 120). Rudyard Kipling today seems to be the epitome of the imperial, and the colonial. Not just his poems and novels, but his life, as told in the autobiographical text from which this fragment is taken, provide deep insights into what we could call the essence of the colonial experience. As a young man, Kipling seems overwhelmed by the plethora of words and the concepts embedded in them, and has, in the extreme, even to smell those words that he would otherwise not be able to grasp. The confusion that emerges out of diversity and the complexity of what is said and what cannot be said may have been unsettling, but it remained a significant working tool for Kipling.

Getting over this crucial confusion and turning diversity into poetry also meant that the words' smells and weight would be understood, or felt, by others who hadn't been where Kipling had been – in Cape Town, Allahabad, Simla. Writing about oneself as part of a giant commotion – intellectually, culturally and geographically – is a common trope in autobiographical texts of the period and requires us to look at the entire body of the writer: ears, tongue, nose, hands, heat and the moisture on the skin, the smell of food on his plate, the ache of limbs after a day's walk. Kipling is extremely precise in his descriptions of sensual perceptions and his bodily reactions to them, and it is striking how his physical presence is there in these texts. We could say almost the same about Flaubert as he writes about his journey to Egypt, about visiting slave markets and brothels.

But what is this really about – is this a matter of genre and a colonial trope? Are the bodies of these Northern visitors to the Orientalized spaces of the colonized parts of the world embedded in a more compelling sensuality? Does the notion of smell and the intensity of feeling convey a sense of uncontrolled diversity, foreignness embedded in the foreign? Is the otherwise closed and controlled body of the European traveller turned into an atavistic body, with smell suddenly becoming a primary sense (Beer, 2000)? In principle, it seems, these texts and their motifs are just exquisite examples of colonial mimesis, '[...] a process and [...] a practice of the colonizers themselves that could bear productively on colonial relations of power' (Ferreira, 2012).

Perhaps we have to change our perspective in order to be able to grasp the ubiquity of such motifs and metaphors. What precisely constitutes these diverse spaces? Who lives in them and how do their inhabitants speak about themselves? What makes these imagined fringes of formerly colonized societies and spaces smelly, messy and confusing? Why do they require their visitors to think about their sensual reactions and bodies in this intense way? Having been interpreted as an expression of Orientalism, sensuality in colonial writing is very much looked at within a Western framework. But perhaps, thinking of the beach and the safari connected to each other through tourist locomotion, we can draw a line from writing in colonial settings to the plurality of speech at these sites.

A Moved Village

Villages that stood in the way of the creation of a national park, a dam or a resort abound. The village of Kiryandongo, just outside Murchison Falls National Park in Uganda, and just a few kilometres from the provincial town of Masindi, was relocated by the British in the 1950s when the national park was implemented. Kiryandongo is claimed to be a Chopi village, owned by a group of people whose ancestors originally lived close to the falls, and located in a region that is characterized by pervasive multilingualism, high linguistic diversity and a very feeble correlation between language and (ethnic) identity concepts. Some Chopi speakers told us they had always lived in 'mixed' villages, then and now, and felt at ease with other people, and therefore that they had easily adapted to the new settling area since the mid-20th century.

This, however, was a very ideologically framed way of dealing with their history, because many Chopi clearly felt that they weren't properly compensated, and because relocation was not the only time that other people had interrupted their lives. During the war of the Lord's Resistance Army against the Ugandan state, villages like Kiryandongo suffered tremendously, with many children being abducted to become soldiers. But experiences of crisis in this area date back centuries. The area was frequently affected by slave raiders and epidemics alike (Doyle, 2006; Médard & Doyle, 2007; Reid, 2002, 2007), and these, as well as local conflict, were responsible for the mobility that characterized the social histories of the Great Lakes regions (Kopytoff, 1987). Conflict and crisis tended to produce new settlements rather than resulting in lengthy civil war, as one of the preferred strategies in conflict resolution was migration. New settlements gained power and safety just through the inclusion of other people. Such strategies have resulted in the conceptualization and practice of multilingualism as a social and political necessity, and as a survival strategy. Current globalized power games and wars have resulted in a high fluidity of migrants and the languages they speak.

Conflict and war resulted in an influx of migrants from almost everywhere. Many villagers claimed that this did not pose any problems to them, as they were welcoming people. The fields around the village were hard to cultivate, they said, with very heavy soil, and manpower was always welcome.

Being aware of the global grasp on agriculture and tourism sites, the villagers clearly noticed that the multilingual repertoires of the area were becoming richer and more diverse. 'Rare' and therefore prestigious languages must have been present for a long time, such as Maltese and Polish, spoken by the inhabitants of a World War II POW camp nearby. Now, however, these languages multiplied. An interesting strategy that mirrors this situation is that the villagers have taken up counting the languages spoken in Kiryandongo. There were 58 languages present in the year 2000, and 72 in 2014. The inhabitants themselves were not counted.

Speaking, Smelling and Settling

Practices such as counting languages illustrate the complex entanglements between the local, the global and the colonial. Nobody could give us any idea of how many languages were spoken in the villages before relocation took place. Language as a separate entity, a countable unit, is hard to think about in such an environment, where the use of several ways of speaking on a daily basis is the norm. Repertoires consist of a variety of ways of speaking, which are used in specific contexts.

Considering its sociolinguistic situation, Chopi is most appropriately seen as one of several ways of speaking – one out of several options speakers are able to choose from. Concepts such as MOTHER TONGUE and NATIVE LANGUAGE (in the sense of Bonfiglio, 2010) are not easily applied. As the community receives increasing numbers of short-term migrants, who live in Kiryandongo only until they can go back or join family members elsewhere, linguistic diversity turns into a multi-layered phenomenon. This is, in principle, very much in line with observations about emerging complex diversities in global tourist spaces.

The resulting relatively new diverse contexts that characterize Kiryandongo today are two-sided. While linguistic practices may not have changed in principle, language diversity has become too complex to simply be incorporated into repertoires. These ways of speaking are therefore used, on the one hand, as 'resources [...] learned in the context of specific life spans, in specific social arenas, with specific tasks, needs and objectives defined, and with specific interlocutors' (Blommaert & Backus, 2011: 21), but they also need to be handled in a way that relates to Northern concepts; they need to be counted, separated from each other, put into order. The presence of foreign NGOs and the accessibility of educational institutions offered by them, the church and the state most certainly invite the appropriation of linguistic diversity to language concepts employed there. Even though the notion of language as a separate entity and discernible unit stands in stark contrast to the actual ways of constructing individual, complex communicative repertoires, it could be filled with meaning. It signifies, in all its coloniality, nothing but modernity, epitomized in the ability to distinguish, to write and to list. As an interpretation of the Northern concepts and practices that came with colonialism, the acclamation of language as a named, countable unit relates Chopi localism to globalized migration and ideas about speech.

The practice of accumulating and discursively presenting linguistic diversity is an interesting concept. Deep with meaning, it relates to a strategy that aims at continuously turning the personal into the communal, the Other into the Self: repertoires are symbolically treated not as something that 'belongs' to a person, but that represents the entire village. Of course, nobody in Kiryandongo speaks 72 languages. But the entirety of the codes represented by the current inhabitants represents what this community 'owns'. While the place and its communality are based on experiences of fragility and precarity, they are symbolically constructed as powerful and enduring by turning personal repertoires into communal diversity. Counting – thereby dividing communicative repertoires and thus LANGUAGE AS A WHOLE – in Kiryandongo makes community, combines different people's ways of speaking and turns them into one huge local sound.

This strategy also works on another level, namely the semiotics of smells, interiorities and memories. Chopi speakers, even though they emphasize their village's ownership of a communal repertoire of 72 languages, expect those who 'bring in' other languages to learn Chopi, and outsiders who learn Chopi can clearly be seen as the norm. A Chopi-speaking foreigner is not received with great excitement, quite in contrast to the ability to speak *other* languages, an attitude which emphasizes a xenophilic language ideology. And therefore, it was not surprising to the people in Kiryandongo that we, as foreigners, were able to speak Kiswahili, or had some knowledge of Chopi, Acholi, and so on. What they didn't appreciate, after enquiring about language use in Cologne, was the German monolingual culture: using just one language on a daily basis in such a setting didn't at all elicit concepts of 'modernity' and 'development' among our interlocutors.

Diversity, however, is not just positively evaluated, but serves as an emblematic identity marker. Ubiquitous motifs and tropes of folklore include family histories as foreigner-inclusion histories, framing the Other, the foreign, the diverse as someone who can and must be incorporated (married, adopted) and thus will be turned into part of the Self.

Linguistic means that help to express, metaphorically, the relevance of turning diverse individuals into part of the communal body, are expressions of smell. Chopi, like a few other Lwoo languages, is a language with a complex terminology of smells and tastes (Storch, 2011). Some examples are:

tík	'rotten, spoilt'
kúur	'perfume, fried food, beer, flowers, baby, smoke'
mît	'ingested food, sweets, mango'
keer	'ginger, chili'
wác	'sour'
ŋwée	'fuel, stench'
kwôk	'germinating food'
tôp	'anything gone bad'
chóò	'repulsive, disgusting'

As the last term illustrates, these concepts are sometimes expressed by making use of different languages, here utilizing a Swahili lexeme. Smell terms are used in discourse in community-making contexts, e.g. beer-drinking, storytelling, resting, etc. Migrants are expected to learn them, and they often do. Most of the forms are used in discourse in a very similar way to interjections (like *kèléelée* 'quiet!'). They are largely absent in narratives, culinary talk, proverbs and heightened language (e.g. prayer, poetry), but surface in dialogical communication. Here, they help to situate social relations:

A   néedi?   'how?'

B   a-tyé   ma-bɛɛ.   may I sit.   kom-á   ò-doko   lím.
    1SG-stay   REL-good   body-POSS.1SG   3SG-become   sweet
    'I'm fine. May I sit? I'm re-energized.'

What is expressed here, in a greeting, but also in comments and remarks during a chat, is the ubiquity of smell in social interaction. And there is no individual way of perceiving social smell, but rather a communal one. Sensuality – in principle very personal and related to the body of the Self – is here turned to the Other, and is framed in the discursive regime of the sensuality of strangeness. This is another part of a huge metaphorical machine that turns single persons, who will perceive smells and taste in individual ways, into part of a huge communal system. There are no personal ways of perceiving odours here, but there are players who are all expected to perceive more or less the same social concept.

This metaphorical complexity is, in principle, just one way of discursively framing community making on the African Frontier. To Igor Kopytoff (1987), the mobility of people, villages, stories and repertoires results out of conflict-solving strategies, such as migration as a reaction to defeat, and the attraction of adherents by those who have started new communities and settlements. Because new conflicts could always result in the break up of a community, it was important to assemble as many followers around oneself as possible. But these people would never switch to new repertoires and fully adopt new concepts of identity; as one always had to reckon with the possibility of being ousted, it was wise to remember clan histories and various ways of speaking, as they remained potentially relevant.

This long-lasting sociopolitical setting produced concepts of standards, which were, as language ideologies, very different from the hegemonic European ideas about national language, linguistic standards and the role of the individual. In spite of their fluidity, these communities required linguistic adaptation, and local standards referred to particular combinations of ways of speaking within repertoires – in Kiryandongo, therefore, Chopi is included in most repertoires, at least to an extent that helps speakers to express the social relationships they form part of.

In the Showcase

STANDARD in Chopi, being conceptualized as and referred to as the composition of repertoires and not the ability to reproduce a particular delimited code, applies to communicative performance and practice – very much in contrast to otherwise dominant linguistic concepts that have to do with structure, the lexicon and literacy. The latter is also claimed by language planners, educationists and politicians in Uganda. Constructing a standard and equipping its imagined speech community with an equally standardized orthography is regarded by most, in principle, as tantamount to development and modernity. But in the case of Chopi, as well as elsewhere in Lwoo-speaking East Africa and into Kenya and Tanzania, a strange twist occurs. As the expansion of Lwoo is of relatively recent date (Atkinson, 1999), Southern Lwoo languages (such as Luo in Kenya, Acholi in Uganda, and so on) are relatively similar to each other (Storch, 2005, 2006). This has fuelled linguistic debates about mother tongue education and language planning, which are basically the only social and political arenas where minority languages could become visible (Namyalo & Nakayiza, 2014). If they are all so similar, why should separate primers, courses and media programmes be launched for all these languages? Wouldn't it suffice in principle, and make more sense economically, to provide the necessary 'tools for development' for the largest and politically most important varieties, such as Acholi, Lango and Kumam?

However, such debates do not just involve economic considerations, but also, and probably much more so, the challenges that rural linguistic diversity poses to linguistics in terms of the conceptualization of LANGUAGE as a fairly open repertoire. The practices of people living in communities such as Kiryandongo are hard to describe by means of traditional linguistics, where the aim is to describe 'a language' in order to be able, for instance, to explore social networking, contact, code-switching, etc. Having received some type of mainstream linguistic training, both project participants from abroad and local authorities engaged with language are asked to produce linguistic generalizations, which can serve as the basis for further linguistic descriptions of Chopi as a mother tongue that belongs to its speakers, thus identifying them.

And this makes sense to these speakers: Uganda's language policy, with English as the official language and Swahili, Luganda and other vernaculars as national languages, is shaped by the colonial idea of separate ethnic languages, an idea which symbolically equips these languages with considerable power:

> The major challenges to the realization of linguistic rights stem from differing and often conflicting ideologies and politics and the general attitudes and perception people hold towards the diversity of languages. Hence there is a need to agitate for prestige language planning. This should, among other things, ensure continuous public language campaigns

and lobbying policy-makers to ensure that language policies, especially those that promote linguistic rights, are translated into blueprints for action other than remaining public relation statements that serve political interests. The same type of planning will also ensure that the public understands why there is a need for people to love, develop and use their languages, not only in their homes but also in public and official domains. Since prestige planning also targets negative attitudes, similar efforts should be made to empower Uganda's indigenous languages to ensure their role in the social, political and economic development of the country. (Namyalo & Nakayiza, 2014: 14)

Successfully showcasing Chopi as a PRESTIGE LANGUAGE, however, has not happened. And there is a specific reason for this: Chopi is not simply a way of speaking as part of liquid repertoires, but a way of speaking that has largely been unmoored from its original site. Of the old Chopi villages nothing is left, apart from the name of an upmarket safari lodge which is located near one of the old settlements: Chobe Lodge,[1] a site of safari-before-beach.

The safari lodge is not a heritage site and not a suitable place for memory culture. It is, however, in all the sensuality it offers (aromatic massages, blue hour cocktails, the refreshing swimming pool), a deeply colonial site that offers various amenities of Western modernity. As such, it epitomizes all that has replaced Chopi cosmopolitanism and all that Chopi cannot be: the colonial concept of the monolingual speaker of an incipiently standardized Indigenous language directly correlates with the creation of borders. These are drawn not only around national parks, but around administered districts, ethnic groups and languages. We do not find just traces of these colonial borders on the linguistic maps of Africa; rather, they continue to underpin practices that are applied by linguists, missionaries and politicians alike.

Consequently, Chopi needs to be fixed as a language on a linguistic map, a classification table, in a development plan and as an ethnic marker in order to obtain a status that would put speakers in the position to claim their own prestige language and thereby their linguistic human rights. However, the rural melting pots whose inhabitants rely on very different concepts of language in order to successfully integrate in rural labour markets and social networks do not provide such anchoring points, and neither does a safari lodge. The ethnic emblematicity correlated with Chopi is limited to just an old rake in a dusty showcase at Kampala's National Museum (Figure 7.1).

Comments on speech, things and words are sometimes mimetic interpretations of sociolinguistic Otherness. Rudyard Kipling, living between and within both the European and the colonized societies and their respective cultural settings, comments on the strange, bewildering words that pour out of these confusing spaces and interactions by almost ingesting them. These words and the sensuality they evoke and require are

Figure 7.1 An emblematic rake, a tool for clearing the ground

framed in an imitation of the exoticism associated with the Other and their languages and narratives. By employing colonial mimesis as a narrative technique, Kipling claims possession over these Other voices which he would otherwise not be able to control.

The speakers of Chopi, and some other Luo languages, appear to do the same thing. Their strategies of *counting* languages as separable, discrete units in places such as Kiryandongo, and claiming the right to *define* standards and political status, have very little in common with the actual ways in which they use language – namely as a very open practice. By counting and defining languages they perfectly imitate the practices of the Other: the monolingual, Western, standard-speaking Other, whom they cannot and will not be in their own linguistic and social environments. These environments require a multilingual – or, better, linguistically open and mobile – speaker. But the experience of coloniality and postcolonialism provokes a strong interest in the interpretation of power represented by statistics, standards, norming and quantifying.

In their daily interactions with those with whom they share their linguistic currencies, however, these counters and commentators handle language in another form: not as a marker of difference, but as a means of enjoying sociability, a playground of the senses and indulgence. Using a large variety of ways of speaking and combining different texts in an intricate manner is not just geared at making oneself understood in a linguistically diverse environment, but signifies openness. Strangers are not excluded by means of complex access rituals or communicative exclusivism, but are invited into communities and encouraged to indulge in verbal pleasure.

It seems as if the multilingual and multi-ethnic settlements of the African Frontier radically challenge our ideas about language and speech communities. Perhaps the inhabitants of these spaces simply use more adequate metaphors than we do. Western thinking about language is still very much based on the notion of the monolingual speaker (as a standard, a norm, a biographical construct) – a concept also reflected in the concept of MULTILINGUALISM, which somehow implies the existence of countable, separate languages stored in a person's mind. Yet words are, like desires, pleasures, the name one carries and the journeys one undertakes, part of a huge social yet transcending wealth, which is displayed in the translocated settlements around the safari destinations as well as on the beach and elsewhere, with different interests in mind and displayed for different reasons. Doesn't this encourage a slight change in our perspective, and a readiness to accept the realities of diversity that leave us as members of a tumultuous, flexible, opportunistic and vital community, and not any longer as distant observers of an endangered, or dangerous, Otherness that needs to be remade into order, that needs to undergo development?

Note

(1) The name Chobe recalls the Lwoo ethnonym *chopé*, as well as the Chobe Lodge in Botswana, which is one of Africa's leading hotels.

**The real STORY! Exciting! Raw!
By the authors of chapters 1-7!!!**

**8 – A new chapter!
On various boundaries**

8 On Various Boundaries

Tourism and Postcolonialism

The tourism industry has been aligned with colonialism in various debates. This might seem unsurprising, as tourism has become one of the world's largest markets in the 21st century and yet continues to create uncomfortable feelings: isn't travelling good for getting to know each other, for learning and overcoming boundaries? Michael Hall and Hazel Tucker (2004), in their seminal volume on *Tourism and Postcolonialism*, suggest that unless travel happens along completely different lines and in fundamentally different ways, tourism continues what plantation colonialism once was simply because the imperial debris produced by colonialism now provides the sights and sites that the tourism industry advertises as romantic and authentic, '[t]ourism […] reinforces and is embedded in postcolonial relationships. Issues of identity, contestation, and representation are increasingly recognised as central to the nature of tourism, particularly given recent reflection on the ethical bases of tourism and tourism studies' (Hall & Tucker, 2004: 2). One aspect of such critique is that the relationship between host and 'guest' (or, rather, client) continues to resemble that between '(post)coloniser and (post)colonised, [that is] primarily seen in the context of the interactions between European nations and the regions and societies they colonised since the onset of European mercantile expansion and imperialism in the fifteenth century, and its subsequent disintegration' (Hall & Tucker, 2004: 3). This makes it obvious where we stand: in a world where 'over 80 per cent of the earth's land surface by the commencement of the First World War in 1914' (Hall & Tucker, 2004: 3) was under European influence.

Following Ashcroft *et al.* (1989), Hall and Tucker identify tropes and concepts in tourism that characterize the conditions out of which postcolonial writing and thinking emerge, and take a closer look at what this might mean for language, text, place and order. Out of the tourism encounter something other than critique appears to emerge: completely owned by foreign interests, tourism is a matter of affirming and enhancing metropolitan capitalistic hegemonies. This observation is important to us because it concerns language and text to a rather large extent. The travel guide and the ubiquitous internet site, as multifaceted as they might be, continue to reflect imperial oppression in the form of 'control over language and text' (Hall & Tucker, 2004: 6). Mainly provided in English or in the six or so colonial languages, tourism texts enforce Orientalism, Othering and binary constructions of Self and Other, and present the different parts of the world as goods for consumption. Hall (1994) shows that even ecotourism discourses and practices are no different in this respect.

What if we didn't even attempt to conceal anything here by attempting academic style? What if we wrote about the impossibility of getting into a relationship with one another that is healing and safe, in a style that reflects the context? What if we wrote a description of communication at a beach as we would – perhaps – write a chapter about the beach in a tourist guidebook? Or, at least, something in between, like a field notebook crossbred with a tourist guide? Something authentic about the Other, always this abstract Other? The Other, in this case, needs to be called a 'Boy', a 'Beach Boy' (Ahmed, 2000; Lawson & Jaworski, 2007).

Worldwide Known Insider Facts

Disruption and cracks in the tourism wallpapers depicting Kenyan beaches – paradise beaches – have emerged through the recurrent news on violence during the elections, terrorism and Ebola, setting free the hidden and repressed images that were so well hidden underneath the beach bliss scenery. Kenya is Africa and Africa is Kenya. The beaches became more affordable destinations with cut-price package offers, while the decline in tourism imposed immense problems on Kenyans working in the tourism sector, who had to fight for each individual tourist and each individual item that could be sold (Akama, 2004; Sindiga, 1999). In order to be prepared for this situation, tourists were asked to pay attention to information on crime and danger (*hatari*) before travelling.

Warnings and travel advice that explicitly mention the unreliability of the beach vendors caused severe problems for everybody working in the beach vending sector. Their main challenge now was to convince the guests of their honesty and good faith in order to benefit from the only business that still remained where they lived. Approaching tourists was now complicated in an environment in which borders were so purposely created and constantly affirmed by all those in control and power. Borders guaranteed peace and silence for the tourists, and meant a substantial loss of sovereignty for the Indigenous communities.

The architecture of these borders was not massive. Low walls, ropes or just lines drawn on the white sands were the demarcations that kept the beach vendors away from tourists and the grounds of their resorts (Figure 8.1). Although it seemed as if the beach belonged to everybody (a pristine and beautiful liminal space, a first encounter zone, a place for love), unspoken yet strict rules told people where to stand, where to be and where not to be. The enclosure of the hotel was safe, as long as the tourists were reminded to keep their distance from what was outside.

Figure 8.1 Indian Ocean, wall and sunbeds

And once we finally got used to the situation of being locked in, being in a safe prison, we didn't feel disturbed by the onlookers any more and seized the hours of peace on sticky mattresses. Alone. Solitary and managed as premium customers until the moment when we stepped out of the enclosure and were met by the waiting Beach Boys for a beach walk. Advice about how to react had already been provided by more experienced tourists, in their blogs and in chatrooms (Insider Tip 8.1). Moreover, the paradisiac hotel grounds abounded with signs and signals: do not go out there, do not step on the grass, do not speak to the Other, do not question this (Figure 8.2). The setting was fixed, with all-you-can-eat buffets and an infinity pool.

The infinity pool, first of all: this symbol of an eternal space, a plane of blue, without waves or birds or plastic garbage. As it extends into the sky and therefore the horizon, and merges with the sea in front of it, the infinity pool suggests endless gazes into the sky, tantamount to seeing heaven, perhaps, the advertisement promises us – the ultimate escape. But the infinity pool is also about new concepts of space, which suggest that there can be more open and flexible spatial planning, a better organization of our immediate built environment that opens up new possibilities. Not a rigid enclosure, like the grounds of the resort, but more dynamic boundaries, more organic, transcending conventional planning. Allmendinger and Haughton (2009) call such new spaces 'soft spaces', and Jay (2018), among others, has suggested that:

> Soft spaces thus reach beyond conventional administrative and institutional boundaries. They take into their scope whatever geographies are suited to the matter in hand, possibly with indeterminate borders and open to expansion. [...] This leads to a picture of marine space-being-planned as lively space, expressing, amongst other things, the sea's materiality. (Jay, 2018: 451)

> **Insider Tip**
>
> It's best to leave all jewelry at home. You can buy beautiful African jewelry made of Maasai beats between 5-20USD
>
> Always negotiate the prices in local shops. The prices you will be told on the beginning is most of the times at least 2 times higher!
>
> Be careful buying wooden crafts. Check the quality of the wood and quality of the paint. Some of the wooden crafts might color your baggage and be eaten by wood parasites within 2 months
>
> Do not carry lot of cash! Do not walk in dark empty streets! Do not show your camera or phone visibly, carry it in a bag. Keep money in body belt under your clothes and be careful with valuables in crowd places!
>
> If you rent a car, under no circumstances stop in the night! No matter what you seen or no matter who was waving on you!
>
> There is malaria (except in Nairobi and high-altitude areas), cholera, typhoid, hepatitis, schistosomisis (this worm-infection is terrible as it sounds), meningitis, HIV, the Rift Valley fever and yellow fever. You can catch Bilharzia if you swim in some lakes, streams or rivers. Our travel advice is to avoid swimming in lakes or walking barefoot in puddles.
>
> Always drink only bottled water!
>
> Be aware of beach boys and try to avoid booking a safari with them as it might end up as a big disappointment.
>
> Source: http://villasdiani.com/kenya-travel-advice-packing-list/

Insider Tip 8.1 Security advice

The sea as a soft site of spatial planning remains the battlefield it became centuries ago, while the infinity pool holds the promise of soft spaces as a tranquil reality.

Swimming above the beach in one of these pools that promise the infinite openness of the world and soft geographies that forever unmake the colonial past is at the same time swimming above the sticky sand, the seaweed and the garbage and those who remove it day by day, as well as those who not even get to remove anything here, who are not needed any longer. The infinity pool, this confined soft space illusion of the Northern South, helps us to pretend that all this is not there. Together with the smooth surface of the clear water, the sound of water in the weir is a soft soundscape.

Figure 8.2 Signs warning about Beach Boys

The infinity pool is where the privilege and status of its inhabitants, rather than softness, are performed, and the only dynamicity besides that of the water lies in its temporary use – in and out, and then lunch – a metonymical sign that indicates social class and particular lifestyles (Littlemore, 2015). Tariq Jazeel (2013), writing on the *Sacred Modernity* of postcolonial geographies, highlights that:

> In tropical modern hospitality architecture, for example, the swimming pool of choice tends to be the 'infinity pool', which offers users views from within the pool that extend seamlessly to lake, reservoir, or sea beyond. Though infinity pools are not unique to this architecture, their frequent deployment by tropical modernists to manufacture infinity is noteworthy in this context. (Jazeel, 2013: 124)

Spatiality in modern tropical architecture, Jazeel argues, is contradictory:

> But the illusion of infinite spatiality is, of course, just that: an illusion. At the Osmund an Ena de Silva's house, Geoffrey Bawa enclosed the plot with a perimeter wall that instantiated a desire for privacy in the growing metropolis. Contemporary tropical modern architecture is indeed still driven by this fundamental contradiction that speaks of the style's classed particularity despite its utopian aspirations. If one of the hallmarks of the architecture is a seamless connection of inside with outside, integral to the production of that aesthetic terrain – particularly within domestic architecture – is the plot's spatial delimitation from a real outside of working-class urban poverty. (Jazeel, 2013: 124)

Looking at poverty needs to be avoided in many ways at the resort as well, sometimes by using the natural boundaries available for segregation, sometimes through language and sometimes by simply running away from it all (Figures 8.3–8.4). Architecture and the social setting into which it is

On Various Boundaries 99

Figure 8.3 Make sure to stay at a hotel that offers beach walks once a day. Armed officers will accompany you and other hotel inmates during a two-hour lasting tour. Beach Boys have to stay in a safety zone, being only verbally able to accompany the group, what they often do by throwing in words like 'kindergarden' or the like. It is unproblematic to ignore them

Figure 8.4 Staying in a place of wet feet guarantees distance from the Beach Boys. The picture is a proof of the possible existence of such a place – not first encounters, but first avoidance. The Beach Boys who have the chance to install themselves between the shore and the entrance of the hotel in order to offer necklaces and dhow tours will catch you if you don't use the trick of running like hell

built relate to one another, as social relationships change with the particular architecture within which they take place (Wacquant, 2017). The openness of pools and chalets, airy and spacious, that at the same time serves segregation, creates highly ambiguous relationships: intimate othering, consumerism directed at bodies and affection.

Linguistic Landscaping

For all tourists who do not stay in the ocean, the 'place of wet feet', it is hilarious to take pictures of running Europeans. It almost seems as if colonial power structures have been reversed. But moving does not change anything. The beach positions us and it positions everyone else around us. It positions, places and scripts: through architecture, classification, separation, money, history, the denial of historical experiences, and the language in which we speak to one another.

The struggle of those Kenyans who had organized their lives in relation to tourism has developed into a struggle for survival. A decreasing number of tourists without a similar decrease in the numbers of tourism workers has led to an imbalance. In contrast with beach vendors in other beach paradises where enough potential customers lie frying on the beach, the Beach Boys at Diani Beach need every single tourist and his or her spending capacity (Berman, 2017; Kibicho, 2009). Owing to this imbalance, individual strategies of performance have developed: attracting and capturing a tourist, walking along the beach and smoothly forcing him or her to buy necklaces, keyholders, coconuts, dhow tours, snorkelling safaris or other kinds of souvenirs and activities. In order to assess a tourist, many Beach Boys have learned to analyse and categorize the groups of tourists with respect to their characteristic behaviour, their nationality and their purchasing behaviour – all necessary knowledge in terms of the way the tourists have to be approached (see Insider Tip 8.2).

To find the right phrase for the tourist is an art that Beach Boys learn at the *Beach Academy*, which stands as a metaphor for the whole beach, on which language is experienced and transmitted. The continuous acquisition of new German, French and Italian phrases, discourse on experience, and constantly testing which phrase excites and captivates the tourists is the basic skill that a Beach Boy needs to start a conversation. The Beach Boys have developed their own kind of performance which includes mainly linguistic strategies in order to play their role in a liminal space within the setting of banal globalization (Thurlow & Jaworski, 2011).

English, as one of Kenya's two official languages, is used in most conversations between Beach Boys and tourists along the beach. But the Beach Boys often use the mother tongue of the tourist, which can be understood as a strategy to attract tourists and to keep conversations going. The Beach Boys make use of a language repertoire that includes German, French and Italian, but also Polish, Czech and Russian and even

> **Germans**
> - need a lot of conversation
> - need to be convinced over days
> - don't buy the same day
> - think it over a lot and get information from other Beach Boys
>
> **British and Italians**
> - seldom book tours at the beach
> - buy small carvings or necklaces to support the Beach Boys
>
> **French**
> - are good customers
> - buy the same day
>
> **Kenyans**
> - are good customers
> - extremely low profit (resident prices)

Insider Tip 8.2 Tourist stereotypes

some words of *kiswisi* 'Swiss German' or *deep kijerumani* 'deep German', which refers to Bavarian dialects (Nassenstein, 2016: 123). The concept of metrolingualism, defined by Otsuji and Pennycook (2010) as a practice of creative use, or mixing, of diverse linguistic codes in mostly urban contexts which overcomes particular social, cultural, political and historical boundaries, as well as ideologies and identity concepts, can be applied to the beach language. Otsuji and Pennycook define the use of metrolingual strategies as an act of inclusion, and at the beach of Diani this is precisely what happens: the inclusion of tourists in business, playing with borders and hierarchies.

Languages are sequestered and performed in a way that enhances positioning strategies: Ethnophaulisms, especially those concerning the Europeans who are present at Diani beach, are found relatively often and mainly at the beginnings of conversations. They serve as discursive starters, evoking a reaction from the tourists, e.g. by laughing about the term or wondering where this knowledge comes from. The frequent use of these terms by the Beach Boys is uncomfortable for many tourists, who later complain about them in their blogs. A beach in a formerly colonized country is a touchy space for many European tourists, where racism and suppression of any kind is to be avoided. The terms presented in Example 1 are never used to address the tourists themselves but rather to make fun at the expense of other Europeans, in order to degrade them and thus to upgrade each tourist's own country and personality. Mock language, if not directed at an interlocutor but at a third person, has comic effects and creates a particular form of hostility (Alim, 2016; Lipski, 2002).

(1) Germans   Kartoffelfresser   'potato eater'
    British   Inselaffen   'island monkeys'
    Italians Spaghettifresser   'spaghetti eater'

(2) Besser den Spatz in der Hand als die Taube auf dem Dach.
    'A bird in the hand is worth two in the bush.'

(3) Es ist noch kein Meister vom Himmel gefallen.
    'No one masters anything without hard work.'

Another way to attract attention is the use of proverbs by many of the Beach Boys, in a convincing way to soften the course of the conversation. The use of proverbs highlights two interesting factors: they are used in the right context at the right time, implying that the speaker has not only learned a German proverb, but also understood its poetics. A proverb such as *Besser den Spatz in der Hand als die Taube auf dem Dach* (Example 2) was a reply in a bargaining situation, when a final price was set for a carving. *Es ist noch kein Meister vom Himmel gefallen* (Example 3), in turn, was the answer to a complaint that a carving was faulty.

This can be seen as subversion of a special kind, where not only the language of the addressee is used, but where a certain speech register is transposed into a different language which would not make use of proverbs in these situations. A reaction of some tourists in this situation was to teach the beach vendors some more proverbs. And when they left they were bidden farewell with the saying *Alles hat ein Ende nur die Wurst hat zwei* 'Everything has an end while the sausage has two'.

Sometimes our tricks fail us. And attention is something that needs to be grabbed in ever new ways. Yet, it felt strange to hear that sentence on the beach: 'I'm not a cannibal. I don't bite!' Being greeted in that way seems to have started in 2012; before then, nobody seemed to have had heard it. 'Oh, no no. We really don't have thoughts like that', was our answer, feeling guilty on behalf of all those Europeans who seemed to have insulted this man. 'That is good', the man continued, 'I only want to sell things to you. Others they steal. I don't. I am an honest person who earns his money without stealing.'

We bought three necklaces and a bottle of aloe vera oil, even adding 500 Kenya Shillings for his daughter, who was accompanying him with a butterfly painted on her face because it was her birthday. How must Europeans have acted on this beach so that this man needed to assure us that he wasn't a cannibal? A pondering silence accompanied us as we left him and his daughter behind. Not for long, of course. We were on Diani Beach where loneliness is a concept of absurdity. A man working as a Beach Boy approached us. 'Don't fear me! I am not a cannibal!'

CHAPTER 9

Hostility on a T-shirt

9 Hostility on a T-Shirt

Items with Language on Them

In tourism, objects with 'language' on them are typically souvenirs, such as cups, decorative objects and T-shirts. Language on T-shirts appears to be particular. It poses a number of questions that emanate from where this language is placed and from where it originates. Is it the language of a place, or rather of a body? To whom does it speak, and of what? And who speaks? And does the T-shirt, this global garment, become local clothing through the words and images that are printed on it? Or doesn't this matter?

A T-shirt is not an easy item to study. It has its indexicalities that evolve out of its colour, material – recycled, eco-friendly or conventional cotton – and price. Not everybody wears it, making it a semiotic tool that signifies membership of a specific group or social sphere. In various contributions on youth culture, it has been addressed as part of style and stylizing practices shared by predominantly urban young people. Most notably, Roland Kiessling and Maarten Mous (2004) mentioned music, hair style and clothing as part of an entire performance of youth language, and this observation has since been repeated by many others. In a sociosemiotic study, Innocent Chiluwa and Esther Ajiboye (2016) suggested that the slogans and mottos printed on T-shirts deserve to be studied more intensively and detached from youth language research. In their work on the discursive pragmatics of T-shirt inscriptions, they treat the messages printed on garments as pragmatic acts through which – in this case – 'Nigerian youths construct their environment and social aspirations' (Chiluwa & Ajiboye, 2016: 436). T-shirt wearers, who in this case study are students, 'take some stances and discursively construct for themselves particular identities by which they position themselves and the Nigerian society they address', the authors observe. To them, 'a T-shirt demonstrates that fashion indeed can speak'.

Nicolas Coupland looks at T-shirt texts in a somewhat similar context: as landscaping in the 'sense of putting language on display in public arenas' in Wales (Coupland, 2010: 79). T-shirts with playfully altered Welsh catchwords on them are 'ironic, creative and deeply coded textual representations of cultural values and antagonisms' (Coupland, 2010: 96).

These perspectives on language worn on the body are informed by the idea that language is fluid and local practice (Otsuji & Pennycook, 2010), and that variation in language practice needs to be researched with a focus on stylistic issues rather than static concepts. Arguing with Penelope Eckert (2012), there is a strong tendency to understand such creative language practices and these forms of linguistic landscaping as ways of positioning oneself as a speaker, of stylizing one's social and communicative appearance and role. This ties in with new perspectives on language as a whole:

> In the move from the first to the third wave of variation studies, the entire view of the relationship between language and society has been reversed. The emphasis on stylistic practice in the third wave places speakers not as passive and stable carriers of dialect, but as stylistic agents, tailoring linguistic styles in ongoing and lifelong projects of self-construction and differentiation. It has become clear that patterns of variation do not simply unfold from the speaker's structural position in a system of production, but that they are part of the active – stylistic – production of social differentiation. (Eckert, 2012: 97f.)

This might make a lot of sense when thinking about language on a Nigerian metropolitan campus, or in Northern settings such as Wales. But what about tourism contexts? Settings that are shaped by messy, conflicting and transgressive practices? At least in terms of items sold and profits made, these are the settings in which the language on T-shirts is really important. They matter for yet another reason, at least. They inscribe what is positioned, already scripted into our bodies, through the ways in which our presences come into existence, through the racialization of class and through the semioticization of the spaces into which we travel. If not for noisy prints and shouting slogans on our bellies, what would remain? Silence, we suppose. Because what can be worn there finds no addressee: the writing on the shirt is not directed at anybody on that beach; it means me/us. But how? And why do we buy and wear this?

Three T-Shirts

Three T-shirts were bought in souvenir shops that almost exclusively catered to European or Northern tourists who visited holiday destinations in the South: Cairo, Zanzibar and Mallorca. The T-shirts were produced there (where they were bought) for customers who were Northern tourists. The shirts were worn at the tourist destinations.

At first, we thought about gossip. A T-shirt is such a leisurely form of dress, so much connected with having fun and being sporty, etc. – this should be a message about the banality of daily life. Moreover, we had recently seen an East African *leso* with a print that was about gossip, so the connection was easily made. But then the T-shirt shown in Figure 9.1 was part of a collection that differed somewhat from the usual fare. The

Figure 9.1 First T-shirt: Cairo

shop that offered these T-shirts also had little pyramids and other such objects, a large collection of T-shirts with prints that depicted the Simpsons as pharaonic Egyptians, and then this.

There was a very visible difference between the usual tourist fare and the designs on display here. On a separate table, shirts that did not easily make sense as stereotyped 'Egyptian' were shown: 'Go to hell' written with Arabic letters; 'This too shall pass' on a pirated *Fajr Army* shirt (Storch & Warnke, 2020); and then this cut tongue: 'Some people need to have their tongue cut.' We were reminded of the failed revolution followed by the military coup in 2013. All over the market, with its colourful yet monotonous souvenirs, there were banners celebrating Abd al-Fattah as-Sisi, Egyptian president since 2014. For four years now, Egyptians had been dealing with increasingly tight control over what they said: by the secret service and the police, with the military everywhere in public spaces. The previous year, we had been to Giza in order to research the multilingual repertoires of tourist guides; we saw very few tourists, but there was a massive secret service presence. Everywhere there were men discreetly dressed in blue suits – sitting at the foot of the great pyramids, strolling along the road that leads to the Sphinx, standing beside the camels that were offered for a ride. The guides and souvenir vendors around were eager to talk to us, about language as well as about its absence – how it had become hard to find tourists to whom one could speak in German, Italian or even English, and how the state controlled and harmed Egyptian citizens. But every time one of these men in blue suits came closer, an entirely different discussion would be performed: talk about souvenirs, prices, tours. Tourist talk. Once the secret service were at a safe distance, the discussion would return to its original topics.

So perhaps the blue shirt with the red tongue on it was about censorship and military rule? About Egypt being a place where citizens had few civil rights? A proof of having visited one of these countries that are ruled by dictators, fundamentally different from Europe, an Oriental country? We wonder whether this T-shirt was made for people who look at the world like this.

Or perhaps there is a different kind of hidden meaning involved here: the T-shirt makers may have been making an ironic statement that is neither kind to the tourists who buy it nor to those it refers to.

The T-shirts in Figure 9.2 were sold in a souvenir emporium in Zanzibar's Stone Town, where people can choose from large amounts of similar things offered in many similar-looking shops. There is very little originality, nothing too outstanding, just recurrent themes of *Hakuna Matata* and Freddie Mercury, mostly produced in China. Yet tourists buy the T-shirts on offer and wear them during their holidays. A particular

Figure 9.2 Second T-shirt: Zanzibar

type of T-shirt stands out: one that is neither about the *Land of Hakuna Matata* nor that portrays humorous messages about life in Tanzania, but that has a very direct way of depicting what is exotic and Oriental in Zanzibar. The picture of a veiled woman is accompanied by the print *BUI BUI ZANZIBAR*, the Swahili word for the veil, and then the non-Swahili word for the island. The Other is a woman wearing a black veil. A *buibui* woman, a woman whose body is invisible, who walks these streets too, at the same time as the tourists, but whose presence refers, it seems, to a different time. Not yet liberated, not yet in hot pants. The tourists who look at this T-shirt are young, and most of the women we see wear very short pants and very short tops. They look inadequate to us, but their wearers do not seem to mind. We think of *The Burka Avenger*, a popular Pakistani cartoon, and whether we can make a connection. The *buibui* woman will not fight for anybody's rights, we assume, but will hover above some unclad legs and very short pants. Look at this Oriental relic, the print seems to say, she still exists, and like the image of a rhino or hippo or shark, one can wear her as a kind of testimony of that particular encounter: with authenticity and the past. There is no superhero depicted on the T-shirt, but a symbol of Orientalism and allochrony (Fabian, 1983; Said, 1978, etc.). And yet ... the *buibui* woman makes a powerful statement. She may be appropriated in an almost unbearable way, but she makes it painfully clear that tourism on Zanzibar is utterly neocolonial, transforming most of the former plantation economy into a tourism economy, tourism already being the island's largest economic sector. Much of this tourism is not in any way interested in the communities on the island and instead is embedded in the global economy, which leaves little and takes much. More recently, communities of the much-visited northern coastal region of Zanzibar have protested against the unwanted effects of tourism, such as drug trafficking, public sex work and improper conduct by the visitors.

This T-shirt, we think, refers to alienation and various forms of violence, which are situated in settings of multitemporality: that what is over is not over. The colonial economies of exploitation are felt, and they have their effects in how we can consume or reject the bodies of Others. And there is reason to believe that the Others know what is happening. More recently, the Zanzibari fashion designer Doreen Mashika has marketed a collection of T-shirts and printed phone cases that bear photographs of Zanzibari people – rural women, Maasai beach workers – who did not seem to welcome those who were staring at them. There was humour in this, too, but also a story about the perfectly audible voices of Others.

At El Arenal, a party tourism destination on the Mediterranean island of Mallorca, the voice of the Other is part of it all: the beach and streets leading to the party locations are not only places where tourists perform transgressions and express their desires for rupture, but also where a very performative encounter with those who provide the party space with the necessary commodities takes place. West African beach vendors, Chinese masseurs, Spanish and Romanian vendors of Romani descent who sell drinks and fruit, and others walk up and down these areas and offer their goods to the – mostly German – tourists. Often, vending involves a certain type of masquerade, for example by Senegalese selling sunglasses, who wear costumes and play with giving names to others and being named. *Helmut* is one of the most frequently used names for both the vendors and their male customers. Female customers are often referred to as *Lady Gaga*, but also as *Monika*, *Gisela*, and so on. Other players participate by mocking or mimicking the Other, such as Nigerian sex workers who address their German clients by simply offering the *ficki ficki, blasi blasi* 'fuck fuck, blow job blow job', and who are in return, mimicked and mocked in the same way by the Germans: *ficki ficki, blasi blasi*.

The T-shirt shown in Figure 9.3, with the printed rejection of all this, has been around for several years now. Its popularity among tourists seems to have increased rapidly with the success of a party hit that was based on an objectification of the *Helmut* persona as an undocumented Senegalese migrant who is yet somehow cool: Senegal, illegal, doesn't matter.

Now, the T-shirt has been offered in a kind of upgraded version: there is not only the rejecting text on the front, but also – headed by the toponym *Mallorca* in Gothic font – *no* in large letters three times, in combination

Figure 9.3 Third T-shirt: Mallorca

with signs that refer to the emblematic businesses of the presumably undocumented migrant workers at the beach. You might not be able to read, but you will understand this, the T-shirt signals. And Mallorca is ours, not yours, so you'd better go. Those who wear these shirts still engage in playful conversations with those whose rejection they embody.

The shops along the *playa* where these T-shirts are sold are almost all run by Indian families. They select motifs to be printed on the T-shirts from a catalogue of a designer and merchant based in Barcelona. About 2000 motifs are available. The range from which the tourists can select a motif contains several dozens of prints, and besides the T-shirt seen here, there are some ten more that seem to be constantly wanted. Most of them are misogynistic, othering and hostile. All of them are considered funny and relevant for inclusion in the party crowd.

The messages printed on these T-shirts that are marketed as tourist commodities are hostile and are part of rejecting speech acts directed at the hosts of the people who might wear these garments. There is a recurrent strategy to construct the host as the Other, as a commodity, Oriental, strange, foreign. Yet those who wear these prints usually do not design them or produce these T-shirts (even though in Mallorca tourists actually do design their own shirts, but these are usually worn only among small groups for special occasions, such as stag parties). They need complicity in making the Other, which we assume are the stakeholders in local as well as global souvenir industries. But they also are complicit: tourists actually wear these T-shirts, at the holiday destination and maybe elsewhere too. There is a strange form of incorporation of this rejected Other going on here: wearing this other skin, sailing under a false sail. The wearer seems to be a friend, one who at least takes notice of the marginalized players around him or her, but then this might turn out differently. The materiality of language (and the language of materiality) as it exists in the context of late capitalist mobilities is a salient part of the global tourism industry, where it is turned onto itself in almost metaphysical ways. These materializations of the Self and Other seem to transgress the boundaries of body as well as place, which makes them semiotically complex and ambiguous. There is no single reading of these transmodal motifs and signs that express privilege and marginalization, closeness and alienation, rejection and complicity at the same time, and that depend on positionality and context to unfold their performative logics of difference. To wear the Other right on your body, to print the Other's name and face and tongue on whatever offers the space for it, what hostility this is!

Language Course

The particular ways in which these T-shirts exhibit language as something specifically connected to a place – for example to the challenges it might pose to its visitors, or to the Otherness the inhabitants of a place represent, or to its celebratedness as a globalized emblematic party space – tend to be understood as a form of language commodification, by turning language into an object and exchange value. The words on the T-shirts do not indicate what they might indicate in other contexts, for example referring to objects and events in which these objects play a role, or to the emotional responses specific encounters may elicit; instead they indicate the meaning of language as something that is immediately and unequivocally connected to economic profit. Language as commodity would be language on a T-shirt, someone's multilingual capacity used as an objective skill (for example of a tourist guide), or a marketed slogan in advertisements, and so on. Monica Heller's work (Heller, 2011, among others) has been crucial in introducing the notion of the commodification of language in late capitalism into contemporary sociolinguistics, and has encouraged numerous contributions on the objectification of emblematic slogans and on speakers' performances and choices, as well as on written language in urban environments (in the sense of linguistic landscapes). Even though many of these contributions have helped to raise awareness among linguists of the fact that language is greatly affected by the extractive and exploitative neocolonial world order in which we live, the debate on the commodification of language – and the term 'commodification' itself – have strangely remained uninterested in a deeper understanding of the ways in which language itself changes through these processes.

In his work on *Social Class in Applied Linguistics*, David Block (2014) problematizes this situation. He argues that the term 'commodity' itself needs to be read in the light of Marxist theory, from where it originally stems, even though many of those sociolinguists who use the term today (including Heller herself) often do not explicitly refer to this context. Block explains further that

> In Marx's work [...], a commodity is a product of human labour. It has value, initially and at the most basic level, for the uses that can be made of it. This value is reflected in how, for example, linen can be used as sheets on a bed, shoes can be used to protect one's feet, a fork is used to eat with and so on. All such objects mediate the satisfaction of basic biological and ideational needs and wants. From commodity production for such basic qualitative value, there is a shift as markets (and market-based economies) arise as sites from which individuals can acquire that which they need or want but which they cannot or simply do not produce themselves. In markets, commodities have exchange value, which means that they can be exchanged for other commodities, as would be the case in a barter economy. Crucially, with market exchange, there is the beginning of a separation between the social relationship of the individual in the production of a commodity and the act of its exchange. The acquirer of a commodity only comes into contact with the finished product but not the socially necessary labour expended by the person who produced it. This is the beginning of a disconnect between the social relationship of production processes and end products. (Block, 2014: 136)

Keeping Marx's interpretation of what is a commodity in mind, Block argues, the idea that language might be commodified becomes questionable. First of all, he observes, a commodity in Marx's sense is always the product of human labour. And language, in all the different ways in which it is acquired, shared and transferred, is hardly the product of human labour. Furthermore, Block observes that, unlike a commodity, language does not owe its usefulness to itself; it never simply serves as an article of something else (such as 'wheat [...] serves as an article of food', Marx, 1904: 34–55, cited in Block, 2014: 137). It has no cost of production that defines its market value, and even an instance of language – such as a proverb or a slogan like *Hakuna Matata* – cannot be measured in value as we cannot, by any means, estimate the costs involved in its production. It is different when we consider the copyright given to individual authors whose text production is in some way measurable and where the product of someone's labour can be valued and marketed. Therefore, T-shirt prints such as *BUI BUI ZANZIBAR* might serve as illustrations of the work of individual designers, whose products are commodities sold at the global tourist market, but who do not commodify Swahili as such.

One could argue, in line with Heller (2011) and others, that the marketing of language skills in the framework of language schools and in the classroom does indeed turn language into a value, costing the amount that one must pay for an evening class, for example. And language courses do indeed serve as indicators for the value a particular language skill might have on the job market. A rising interest in Chinese in many parts of Africa is reflected by an increase in the numbers of Chinese language courses at universities, private language schools and on online forums. But the value of Chinese is not the value of the language itself, but of the labour provided by the administration and teaching staff of those institutions, the materials produced by authors of language course books, the costs involved in printing and photocopying. Turning the gaze to the tourist market once more and the ways in which language courses and learning are represented by the souvenir industry, Block's point might become even more obvious. Souvenir shops in Kenya and Tanzania sell large varieties of T-shirts with language course materials printed on them. As a genre of its own, the language course T-shirt might insinuate even more than other T-shirts, such as those presented above, the commodification of language. The linguistic material on them is usually combined with drawings of the objects denoted by the words in the language in question, and everything is arranged in a symmetrical, orderly way on the front of the garment, as if the systematic process and normative practice of the language course is to be embodied by the wearer.

An example is a child's T-shirt purchased in Diani, Kenya. Made of eco-friendly cotton (according to the label), the object has been advertised as *cool wear *original designs *Africa made. The T-shirt is of an olive-brownish colour that evokes memories of safari advertisement designs.

While the back is devoid of any print, the front bears a print that extends from just under the neckline down to the bottom seam. On top, in multi-coloured capital letters written in a 'playful' font, it says *KISWAHILI FOR YOU*. Underneath, we are provided with a vocabulary[1]:

(1) JAMBO   – HI
     RAFIKI   – FRIEND
     NZURI   – GOOD
     ASANTE   – THANK YOU
     WANYAMA   – ANIMALS
     SIJUI   – I DON'T KNOW
     HAKUNA MATATA – NO PROBLEM
     CHAKULA   – FOOD
     KWAHERI   – GOOD BYE

To the left and right of the wordlist are drawings of a Maasai warrior, a lion, a chicken leg, a sun, a hand waving and so on. A small colourful braid that imitates African prints on kangas can be seen underneath.

A T-shirt in an adult size was bought in the same tourist environment in Kenya and resembles the first T-shirt in terms of its olive colour, again suggesting safari contexts. The print is arranged in a similar way, with a heading underneath the neckline, vocabulary on the belly, images of objects to left and right, and a decorative 'African' element underneath. The vocabulary offered by the T-shirt is much more elaborate and not just written in capital letters. Moreover, the images are included in the language course more explicitly. The heading is bilingual – *LEARN SWAHILI Jifunze Kiswahili*. Underneath we read:

(2) Hello   Jambo
     How are you?   Habari gani?
     Fine   Mzuri sana
     Welcome   Karibu
     Goodbye   Kwaheri
     Thank you   Asante
     No problem   Hakuna matata
     Friend   Rafiki
     Enough   Inatosha
     Shop   Duka
     Market   Soko
     Food   Chakula
     Tea   Chai
     Spoon   Kijiko
     Coffee   Kahawa
     Beer   Pombe
     How much money? Bei gani?
     Slowly   Polepole
     Small   Kidogo
     Big   Kubwa
     Please   Tafadhali

Underneath, khaki-coloured letters say WELCOME KARIBU, and to the left there are drawings of animals with explanations underneath each of them: *Lion – Simba, Zebra – Punda Milia, Elephant – Ndovu, Cheetah – Duma*. To the right there are *Leopard – Chui, Giraffe – Twiga, Buffalo – Nyati, Hippo – Kiboko*.

A third T-shirt of this type was bought in a tourist emporium in Stone Town in Zanzibar. It is a thin white shirt without sleeves, with no print on the back and a symmetrically arranged vocabulary on the front. The heading reads ZANZIBAR LEARN SWAHILI, and underneath an extensive vocabulary is offered:

(3)
Hello	Jambo
How are you?	Habari?
Thank You	Asante Sana
Welcome	Karibu
No worries	Hakuna Matata
OK	Sawa Sawa
Slowly Slowly	Pole Pole
Please	Tafadhali
People	Watu
Hot	Moto
Water	Maji
Tea	Chai
Coffee	Kahawa
Sugar	Sukari
Bread	Mkate
Food	Chakula
Fish	Samaki
Meat	Nyama
Shop	Duka
Money	Pesa/Fedha
Market	Sokoni
Small	Ndogo
Big	Kubwa
Day	Asubuhi
Afternoon	Alasiri
Evening	Jioni
Night	Usiku
Today	Leo
Yesterday	Jana
Tomorrow	Kesho
Goodbye	Kwaheri

To the left, images of animals are labelled as *Rhinoceros ... Faru, Cheetah ... Chui, Elephant ... Tembo, Ostrich ... Mbuni, Hippopotamus ... kiboko*, and to the right as *Lion ... Simba, Zebra ... Pundamilia, Giraffe*

... *Twiga, Impala ... Swala, Buffalo ... Nyati.* There is no decorative element underneath the vocabulary.

Similar motifs are offered as prints on other objects that can be used in everyday routines. A placemat made of laminated cardboard has a zebra stripe design and images of zebras as a background, on which the heading *I ♥ Kenya LEARN SWAHILI* has been printed. Below, there is a wordlist that is remarkable insofar as to some extent it attempts to makes sense of the connection of the object with the meal:

(4) Hello   Jambo
     How are you?   Habari Gani?
     Fine Beautiful   Mzuri Sana
     Welcome   Karibu
     Good Bye   Kwaheri
     Thank you   Asante
     No Problem   Hakuna Matata
     Friend   Rafiki
     Food   Chakula
     Tea   Chai
     Coffee   Kahawa
     Beer   Pombe
     How Much Money   Bei Gani
     Slowly   Pole Pole
     Please   Tafadhali

To the left and right of the wordlist, there are images of animals and maps of Africa and Kenya.

A wall hanger and a fridge magnet, also bought in Kenya (Diani), offer similar language material. The wordlist is printed on a space that resembles an ancient scroll and reads:

(5) Jambo   – Hello
     Kwaheri – Goodbye
     Hakuna   – No
     Matata   – Problem
     Asante   – Thank you
     Karibu   – Welcome
     Pombe   – Beer

On top of it, the heading says *LEARN SWAHILI*, and underneath *KENYA* has been printed on the scroll. To left and right a Maasai woman and a Moran can be seen.

A coffee mug from Kenya (Diani) asks us to *Jifunze Kiswahili – Learn Swahili* and bears vocabulary and illustrations of the words to be learned, arranged in a slightly messy fashion. The vocabulary is thematically less consistent, as if the designer wanted to make reference to the ways in which chats over a cup of coffee meander from here to there:

(6) Kalamu   – Pen
     Kitabu   – Book
     Karatasi – Paper
     Unga   – Flour
     Mkate   – Bread
     Asante   – Thank you
     Maziwa   – Milk
     Habari   – Hello
     Mzuri   – Fine
     Maji   – Water
     Kahawa   – Coffee
     Hatari   – Danger
     Simu   – Phone
     Chai   – Tea
     Dawa   – Medicine
     Tabasamu – Smile
     Matunda – Fruit
     Gari   – Car
     Baba   – Father
     Mama   – Mother

What is being marketed as a commodity here are objects that have been produced by local designers or Chinese (or other foreign) companies. The language printed on these objects always consists of more or less the same words and phrases, and it is always Swahili. As in the context of *Hakuna Matata*, Swahili is taken here to represent Kenya, Tanzania, Zanzibar, Africa. Never do any words in Maa or Gikuyu or any other language spoken at the sites of tourism appear on these objects – Swahili vocabularies and slogans adorn these things in the same ways as the images of animals seen on safari trips do, or the decoration and colour associated with 'African' aesthetics does. All this is not really 'Swahili', as a language or a practice, but an extremely reduced and emblematically framed representation of what a tourist might, after a short holiday, consider to be an 'African language'. The words taught to tourists by guides and souvenir vendors reappear on these objects and remind us that the tourism industry involves different forms of labour. The transfer of emblematic vocabulary in order to make the safari or guided tour more interesting and local, in the sense of more 'African', is part of the work that is done. But this work does not consist of the production of the language itself, nor do the designers of souvenirs produce Swahili; they produce representations of an experience of a Swahili-speaking tourist site.

These reduced and simplified representations of Swahili have their earlier counterparts. As observed by Judith Irvine (2008), among others, the beginnings of African linguistics are characterized by representations of African languages as languages that lack any complexity. Data on African languages, Irvine shows, were often collected and produced with the help of people who had left their homes a long time ago, had been kidnapped and enslaved, or were met by the early missionary linguists in one of the many refugee camps that existed in various parts of the continent after the abolition of the slave trade. Irvine argues that there is a close connection between imperialism and the lacking representations of African languages well into the era of decolonization. The texts on which our knowledge of African languages is based are problematic to no small extent, and this should be carefully reflected on:

> These conditions of linguistic work tended to result in representations of African languages in reduced versions, lacking in social deixis other than pronouns and kinship terms, and simplified in syntax. (Irvine, 2008: 331)

The decoration of souvenirs suggests that learning a language such as Swahili is still the appropriation of the reduced and simplified language of the Other, as if the Other was not able to express anything complex, as if the vocabulary on the T-shirt was just as good as the *buibui*-wearing woman from Zanzibar for representing the exotic and Oriental Otherness of the colonial place and its inhabitants.

Language printed onto a commodity undergoes two processes of alienation, it seems – the disconnection between the person who produces the commodity and the person who buys it in a market economy, and the disconnecting relationship between the (presumed) producer's way of speaking and the customer's estimation and interpretation of it. Power imbalances that are nothing more and nothing less than imperial debris suddenly become painfully discernible through the marketing of commodities to which the language and the image of the Other are attached. And even if these T-shirts and mugs were considered too banal to be looked at in any detail, the effects of the ruinous representations of Othered language remain discernible. As Cronopio (2019) argues, discourse on and at the sites of tourism – in this case backpacker tourism in Morocco – can be analysed as predominantly interested in the expression of egocentric perspectives, where the people who inhabit the places visited by the travellers are reduced to decoration and accessories that provide some sense of authenticity. The travel books that are found in backpacker hostels, hotel rooms and libraries, as well as in the bags of the tourists themselves, provide similar insights. The reduction of a language to emblematic decoration pervades guides and travel books, where literacy pertaining to the Other very often constructs Orientalist images of stereotyped people who have no real language of their own, but only dialects and idioms. These figured Others need someone to give them a voice, it seems, a voice rather than words.

In James Penhaligon's (2011) *Speak Swahili, Dammit!*, an autobiographical text on the author's childhood in Tanzania from 1966 onwards, language is provided as snippets. These snippets are easy to identify as they are always printed in italics. They stage authenticity, give credibility, raise curiosity. This could have been said about many other books that are sold in the bookshops at airports and the travel book selection of online booksellers, but Penhaligon's book has such a good title that we have chosen his volume and not anybody else's. Open the book randomly. 'Fifteen', pages 169–188. The chapter's italic version reads as follows:

(*want*; English; emphasis)

Watu 'Samaki muzuri' 'Sitaki' Mzungu? 'Sitaki' 'Usi pigana' 'Nami kukimbia' 'Apana' 'Kwa nini?' totos watu askaris schutztruppe Afrika Korps Kriegsmarine askari KAR askari KAR Central African Rifles Kimbo

(*that*; English; emphasis)

KAR KAR askaris Schutztruppe KAR KAR watu KAR askari KAR Schutztruppe askaris askaris bibi Schutztruppe askari bibi Whitecap askaris rafikis 'kushoto, kulia, kushoto, kulia' Kushoto Kulia askari 'Hawa' kak-ak-ak-ak-a Plop-plop-plop askari Mzungu kunya askari askari nyokas ka-boom Mzungo ndogo askari Brrr-aa-a-ppp Anakufa kabisa banzaai banzaai 'Asante, ndugu ndogo' hakiri ugali mboga mboros jigi-jigi Two-Mboros Three-Mboros Four-Mboros Kwa kibiriti ya kupata matiti, katika Dodoma kupata kuma lakini kwa rungu kupata mkundu! Rungu novus actus interveniens mchawi mwamba watu ng'ombe mzee watu jinni djinn mchawi wasikini masikini masikini masikini 'Jambo' mwamba 'Nini Mzungu?' wasikini mchawi dawa 'Ku-ulisa' 'Aw-omba' hekima hekima 'Uta rudi'

(*will*; English; emphasis)

Mchawi masikini masikini

(*in*; English; emphasis)

Wazungu bibi jigi-jigi mutu jigi-jigi jigi-jigied bibi Whitecap kanzu fez Harley Davidson Whitecap Harley Davidson

(*him*; English; emphasis)

Mbuia Wazungus Whitecaps Wazungu bibi shamba Wazungu 'Morning Please Don't Come' Mzungu Whitecap shamba Timbuktu makuti maninga pièce de la résistance Club Timbuktu

(*spirit*; English; emphasis)

Club Timbuktu watu Wazungu SS Orion Orion Warabu Swahilis Chagga bibi kuja hapa, leta Watoto Mzungu 'Leta meno ya tembo' watu Timbuktu Mzungu Whitecap watu kabila Johnny Walker bibi bibi bibi.

The words do what they have to do, in this commodity-book. They put into the text what it needs in order to convey the weight of experience and lived-through reality. German words that are about war, Japanese words that are about the battle cries of the Kamikaze pilots, Swahili that is about witchcraft and obscenity, poverty and women. And then there are the different names for different people: *Johnny Walker tribe*. On the back cover of the book, recommendations can be found. The Executive Director of the KAMUSI PROJECT INTERNATIONAL, The Living Internet Swahili Dictionary, Switzerland writes: 'What a compelling window into a unique part of the Tanzanian experience ... a different world, illuminated from a fascinating perspective.'

Peter Moore's (2002a) *No Shitting in the Toilet*, a 'travel guide for when you've really lost it', offers autobiographical data as well, but from the perspective of a traveller. Swahili is explained where it is most saliently encountered by travellers, namely in emblematic soundscapes and foodways:

TOP TEN TRAVELING TUNES

4. 'Jambo Bwana': The Kenyan Safari Band
To me – and to anyone else who's heard of it – this song is Kenya. Contrary to popular opinion, it isn't the happy, carefree nature of this tune that endears it to folk. Nor is it the fact that it uses the only two Swahili phrases most people ever remember – *jambo bwana* and *mzuri sana*. No I'm afraid it's rather more perverse than that. It's because you hear the bloody thing so many times
(From *This is the Kenyan Safari Band*, produced by some crappy pirate tape outfit on Tom Mboya Street, Nairobi) (Moore, 2002a: 204)

Matoke
Mashed plantain bananas and maize. Second only to ugali as the most unpalatable meal on the planet. (Moore, 2002a: 253)

Ugali
A tasteless mixture of maize and water that leaves you feeling leaden and queasy. Unfortunately, it will be your staple diet in Africa. (Moore, 2002a: 256)

We know this trope. *Ugali*, like Swahili, represents Africa.

In Moore's book about his trip from Cape Town to Cairo, *Swahili for the Broken-Hearted* (2002b), this becomes even clearer as we browse through the chapters. Each of them, regardless of which experience, city or country it describes, begins with a Swahili proverb. All of these proverbs are translated into English, which makes this book a bit similar to the language course souvenir T-shirt. A selection of proverbs reads as follows:

Baada ya dhiki, faraja.   (Prologue)
After hardship comes relief.

Mwenda mbio hujikwaa dole.   (Cape Town, South Africa)
A person in too much of a hurry stubs his toe.

Kila ndege huruka kwa bawa lake. (The Garden Route, South Africa)
Every bird flies with its own wings.

Shoka lisilo mpini halichanji kuni. (Lesotho)
An axe without a handle does not split firewood.

There are 21 chapters, one on each destination or stopover, each with a proverb of its own. The chapter on Zanzibar begins with *Asiye na mengi, ana machache* 'Even he who has not many troubles has a few'. Apart from the trans-African proverbs, there do not seem to be any Swahili words in the book, even in the Zanzibar chapter. Instead, the words printed in italics in this chapter refer to a Monty Python film and to classic English literature *(Life of Brian, Wuthering Heights, Sons and Lovers, Tess of the d'Urbervilles, A Pair of Blue Eyes)*.

Novels for proverbs, Swahili words for English words, a barter trade of language on a market of desires and phantasies, where anything gets sold. The texts sold in the souvenir shops do what we might expect them to do, as they continue drawing images of an Other who has no Self, of destinations frozen in exoticism, of language that says nothing. Linguists, who produce much more complete and inclusive representations of language, usually do not see this as a problem. But what is actually becoming visible in this market of dreams is that the imperial formations that govern the ways in which language – or rather speech – is taken control of and made alien to the speakers remain firmly in place.

Note

(1) All examples are transcribed from the original objects, faithfully reproducing font and format.

Chapter 10

MOVIES ON SEX TOURISM THAT YOU SHOULDN'T MISS

10 Movies on Sex Tourism that You Shouldn't Miss

Competition, Image, Desire

Tourism is not only a form of short-term migration that involves excessive consumerism in a colonial setting, where the paying client is served in the way in which colonial elites were served, by servants whose mobilities tend to be far more limited, but it also bears in its consumerist excess the urge to compete – for the most authentic experience, the best location, dinner, beach, cocktail, trail, the most exciting experience, and so on. One of the iconic formats of celebrating competition and consumerism is that of travel guidebooks and travelogues that explicitly offer information about only the best, the top, the must. A hundred places you must see, ten roads you need to travel, the top five of them all. This chapter offers an overview of the best films on the absolute, namely sex. The best films that we discuss are the best films because we saw something in them, found them emblematic or convincing or strikingly banal. They are all films that were produced in the past few years, recent cultural productions that reflect the acceleration of travel and consumption, of extraction and construction.

Nothing in these films is new or novel in any way. Just prior to the time of decolonization, Frantz Fanon wrote about the effects of imperialism on bodies and economies, pointing out that the new middle classes of the periphery, colony, post-colony, Global South, etc., differ from the Western middle classes insofar as they contribute not simply to exploitation and extraction *somewhere* in order to benefit from the resulting profit *ultimately* and *directly*, but that they also exploit their immediate environment in order to provide profit making for other people, for the metropolitan, globally northern middle classes and for the former colonizers:

> The national bourgeoisie organizes centers of rest and relaxation and pleasure resorts to meet the wishes of the Western bourgeoisie. Such activity is given the name of tourism, and for the occasion will be built up as a national industry [...]. The casinos of Havana and of Mexico, the beaches of Rio, the little Brazilian and Mexican girls, the half-breed

thirteen-year-olds, the ports of Acapulco and Copacabana – all these are the stigma of this depravation of the national middle class [... who] will have nothing better to do than to take on the role of manager of Western enterprise, and it will in fact set up its country as the brothel of Europe. (Fanon, 2004: 101f.)

The desire for the consumption of bodies in a physically exploited colonial world has its particular history in popular culture and the entertainment industry. It does not simply emerge out of need or opportunity, but relies on images that are already burnt into our retinas. Tales about Oriental and exotic men who seduce European women were among the first box office hits of the emerging film industry in Hollywood. The silent movie *The Sheik* (Melford, 1921), starring Rudolph Valentino, is among the earliest successful films of this genre, which to a large extent was as much about colonial propaganda as it was about desire and repression (Storch, 2020). Its heroine, Diana, a young woman portrayed as a 'typical' member of colonial British elites, does what women were not supposed to do then: she decides to ride away into the desert to discover whatever there is to discover, on horseback and with a guide and porters by her side. In the desert, however, she falls into the hands of the Sheikh. The Sheikh is young and clean-shaven, a smoker of cigarettes and wearer of wrist watches. The night before, he had climbed into and out of Diana's apartment, singing a Shalimar song underneath her balcony. Now he is smitten by her white hands which, he says, have rosy fingertips, and he intends to have sex with her. After weeks of living with him, she falls in love and – after various adventures involving much horse riding – remains at his side (while he lies in a bed recovering from a fight). The images of the colonial Other, who is characterized by his violent masculinity and boyish smile alike – the infantilized colonial Other who is sexually menacing – have continued to shape fantasies and imaginations of how exoticized and Orientalized men look and what they do for metropolitan White women: fulfil them sexually and liberate them from the tedious tasks of everyday life. In her biography of Rudolph Valentino, Emily Leider (2003) writes:

> [... T]his abduction scene in particular made the hearts of women – not all, but many – go pitter patter. The image of a young maiden being literally swept away by a dark, handsome, and muscular man wild with desire for *her*, the one woman he wants more than any other, set off female hormonal switches. (Leider, 2003: 165)

Female audiences, she argues, were made to sense that there might be something far more interesting waiting for them than household work and monotonous labour. An adventurous life full of opportunities for self-actualization was, however, not to be found where they lived but in the exotic lands of Orientalist phantasies, which by that time had utterly real counterparts in the colonized world. And like the colonized subjects in the Global South, the women in these texts had to turn into obedient subalterns:

> No longer compelled to stay with her Sheik, she elects to stay, having become a slave to love. Broken like a wild Arabian horse that once roamed free, she accepts the harness he offers and finds her maternal, tender side. By her changed behavior she ratifies a hardly subtle credo, on the Sheik shared with Sigmund Freud and Edith Maude Hull [the author of the novel on the Sheik]: only by yielding, by shedding their boyish trousers, handing over their revolvers, and becoming nurturers can females realize themselves as fulfilled women. (Leider, 2003: 164)

And once it emerges that the Sheikh is actually an orphaned European raised by Arabs, order is fully restored. Love turns out to be worthwhile and legal, and the curtain falls.

The production of popular cultural artefacts that promise similar gifts is enormous, giving way to tourism industries that promise fulfilment of these dreams. Productions such as *Eat Pray Love* (Murphy, 2010) have greatly contributed to an increase in tourism to specific countries or regions (such as Bali) and resonate in a large corpus of texts. As a form of cultural mobility, such images and stories trigger specific cultural practices, social strategies, emotional reactions and economical processes rather than simply reflecting them (White, 1973). We have not considered the 111 BEST MOVIES ON SEX TOURISM THAT YOU MUST NOT MISS for this chapter, but only a few, which we felt was enough. Watching them again and again, we found ourselves consuming film after film in which body after body was consumed.

We assume that without telling their monotonous stories about sex tourism, these films, all based in countries of the Global South, would not get much attention. Just as books about Africa in which no 'scandalous' love story occurs would not sell well, these films receive attention only when a dark-skinned person depicts a dependent lover giving up his or her ego, language or behaviour as a result of unequally distributed power relations. The films all deal with what Skinner and Theodossopoulos (2011: 15) call cultural-*cum*-sexual tourism, which can be seen as the extreme result of 'stranger fetishism' (Ahmed, 2000: 114), where postmodern consumption constructs the stranger as an impossible (or unreachable) figure. Here the consumer places himself or herself in the position of the stranger with the help of commodities, as if it were possible through certain products, e.g. coconut products. Ahmed assumes that consuming a stranger involves 'a transformation in the subject who consumes' (Ahmed, 2000: 115). This kind of transformation is the focus of all five movies that we have analysed: a transformation into the exotic Other through the consumption of the exotic's body. The sexually connoted exoticization of difference is, in the sense of bell hooks (1992), directly given in scenes of some of the movies, where the exotic Other is 'eaten' or bought in relation to fine food.

It is an astonishing experience to encounter the plethora of novels in which a North-South love relationship is the only topic, accompanied by the knowledge thereby gained of the environment, culture, politics and other aspects of the lover's home country. The bodies of the lovers are objectified and brought into the same context as other stereotypical authentic souvenirs which are carried back home after travel abroad. The dichotomy between 'tradition and modernity' is, on the one hand, the reason why the novels on which the films are based could be written by their female authors at all, and on the other hand probably the only reason why the novels have ever been read. The encounter with the unexpected Other is the essential aspect and is nothing but a stereotype of the above-mentioned dichotomy, expressed through contrasts between 'beauty and cruelty' or 'pleasure and danger' (Meiu, 2011: 99; Skinner & Theodossopoulos, 2011: 16). The ubiquitous paraphrasing of female sex tourism into terms like 'love' or 'relationship' has given rise to this socially accepted form of travelling abroad, culminating in travel guides like *Romance on the Road: Traveling Women Who Love Foreign Men* (Belliveau, 2006).

The desire for authenticity and wildness, combined with the plea for a life back at the roots with a focus on essential things, made the imaginaries of many of the novels possible. The idea behind the target project of consciously selecting an exotic Other is found not only in tourism and not only by females but also in many other fields of mass culture where pleasure in the acknowledgment and enjoyment of racial difference is found (hooks, 1992: 366).

> Getting a bit of the Other, in this case engaging in sexual encounters with non-white females, was considered a ritual of transcendence, a movement out into a world of difference that would transform, an acceptable rite of passage. The direct objective was not simply to sexually possess the Other; it was to be changed in some way by the encounter. (hooks, 1992: 367)

The films that we discuss are *Sand Dollars* (2015), *Patong Girl* (2014), *Die weiße Maasai* ('The White Masai', 2005) and *Paradies: Liebe* ('Paradise: Love', 2012). All these films are well known and some of them have gained considerable critical recognition. *Patong Girl* received the Grimme Prize 2016 in the category 'Fiction/Special'. In addition, actress Aisawanya Areyawattana, who played the transgender girlfriend, received the Best Trans-Performance award at the 2015 Transgender Film Festival. Nina Hoss, the leading actress in *Die weiße Maasai*, received the Bavarian Film Prize. *Paradies: Liebe* was awarded Best Film Production at the Austrian Film Awards, as well as winning the categories of Best Director and Best Actress (Margarethe Tiesel).[1]

Films

The film *Die weiße Massai* and the novel on which it is based contributed considerably to the increase in German and Swiss tourists – especially women – to Kenya, especially its beaches, in the 1990s and early 2000s. The novel by Corinne Hoffmann and the subsequent film adaptation met diverging reviews, although the positive reviews greatly outweighed the negative. The works became a bestseller and a cinema hit, respectively. Differentiated analyses and comparisons of the book and the film soon met with scientific interest (see Maurer, 2015; Perić, 2013). Perić (2013) offers a comparison of the storyline in the book and the film, with a focus on the scenes where the film varies from the book: variation is often used to insert scenes that indicate discrimination against the Kenyan lover, such as the scene in which the woman wants to invite him for a drink in the hotel bar, but he is not let in by the bouncers. In the book, this scene did not take place (Perić, 2013: 16). Another scene shows both of them in the immigration office in Nairobi, where the man is not allowed in because of his clothes (Perić, 2013: 22). Interestingly, in both places, it is not Europeans who deny access, but rather the Kenyan authorities denying their own compatriot. Thus, the commitment of the woman against discrimination comes in without having to bring up the perspective of racism. In novels such as *Die weiße Maasai* and others (e.g. Hachfeld-Tapukai, 2004; Wiszowaty, 2009), presuppositions such as socioeconomic inequalities are depicted as cultural differences or material poverty and are figured as a source of authenticity (Meiu, 2011: 103).

Most reviews of *Sand Dollars*, which are consistently positive, substantiate the film's success through the masterly acting of Geraldine Chaplin and Yanet Mojica, who played the two leading roles (e.g. *New York Times*,[2] *Variety*[3]). The subject of the film, which focuses on both sex work and unequal power relations, is treated in an astonishingly uncritical way. A romanticization of these issues is also found in the other films that we analyse here, with the exception of *Paradies: Liebe*. It seems that the better the cast and the better the acting, the more obscure is the aspect of sex work or prostitution. The most critical notion that could be found in the reviews for *Sand Dollars* was the issue of mutual exploitation.

In contrast with these films, with their exploitational backgrounds, is *Patong Girl*, a film where the overprotectiveness of mothers, transgenderism and gender inequality constitute the core statement. Reviews of *Patong Girl* are not found in online newspapers, but rather on film review websites. In contrast to the other films, *Patong Girl* has the most unveiled linguistic statements, which make it easy for the viewer to quickly realize clichés and stereotypes.

Paradies: Liebe stands out from the other films, especially in terms of direction. Ulrich Seidl, the Austrian director, is famous for bringing performances of reality to the screen, unadorned and blunt. In contrast to *Die weiße Maasai*, for which, in casting for the male lead role, about 150 Kenyan men were seen but the decision then went in favour of the Burkinabe-born French actor Jacky Ido, Ulrich Seidl hired a Kenyan amateur actor for the lead role. *Paradies: Liebe* captivates through its critical illumination of the unromanticized sex relationship, and the protagonists' mutually depressing influences. Language is used to shock and embarrass rather than to romanticize and veil.

Films of this kind helped to construct exploitative encounters as ethical projects: in a pre-modern Africa in need of Europe, in spaces that are constructed as spaces full of nature and not yet as landscapes, with people in them who speak strange languages, unable to communicate with modernity, unable to make decisions. And so the tourist makes decisions, speaks, stays, copes, writes a book. Testimonials of female writers who describe their intimate experiences in an African country have soared during the past decades:

> Literary works that devote themselves to such love adventures benefit from a Western yearning for exoticism, the desire for forbidden, unconscious instincts of power and/or the need for compensation for powerlessness. Interestingly, the activation of this pattern of emotion works only in a constellation of sex and race, and specifically if the woman is white but the man is not. The search for love in the stranger and the intoxication in and through the exotic are basically embodied by women, while European men pose as conquerors, researchers and traders.[4] (Luttmann, 2007: n.p.)

All the films share the idea of showing the unequally distributed power and resources of the European sex tourists and the local lovers in order to transfer this interpersonal inequality into an international one, which automatically leaves the spectator with the everlasting argument that the support of these exploited lovers helps them more than it harms them. This dilemma has been voiced by Foucault and is felt by many people in the West, where the 'primitive' Other is the target for Westerners' own selfish pleasure. bell hooks points out that:

> It is precisely that longing for the pleasure that has led the white west to sustain a romantic fantasy of the 'primitive' and the concrete search for a real primitive paradise, whether that location be a country or a body, a dark continent or dark flesh, perceived as the perfect embodiment of that possibility. (hooks, 1992: 370)

That this kind of prostitution is not labelled as prostitution has a merely economic aspect in the media, as well as in the real encounters of *romantic tourism*. Dealing with sex tourism in the print and film media far exceeds the idea of how prostitution should be handled in public. Acceptance or even advocacy of this type of exploitation by European women of African, Caribbean or Thai men would never take place in reversed gender roles.

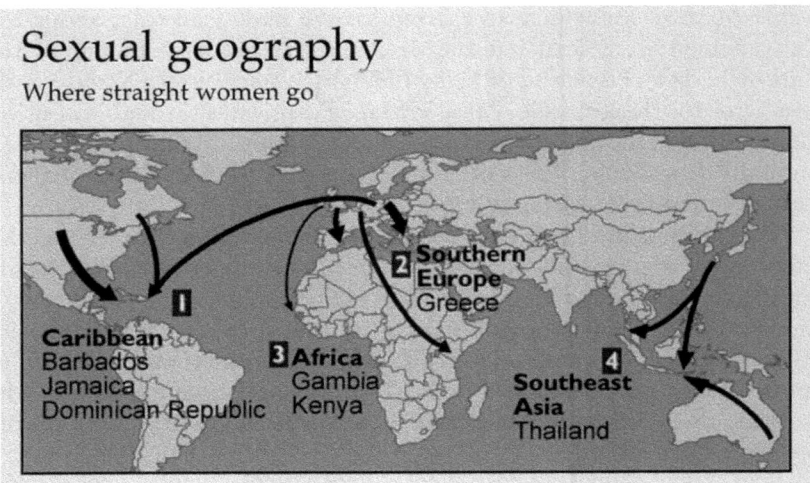

Figure 10.1 Sexual geography

The films that exhibit these problematic issues are not lauded for their critical approach, but because of their open treatment of women's power and a new equality in sex tourism, as if sex tourism has become ethically acceptable. *Romance on the Road: Traveling Women who Love Foreign Men* by Jeannette Belliveau shows exactly this tendency, which is most obvious in the films: exploitation of men by women is interesting and financially worthwhile for the film industry, whereas the same situation with reversed gender roles is disgusting and better swept under the carpet.

The films that we have selected in a random choice all take place in settings that are favourite destinations for female sex tourism: the Dominican Republic (*Sand Dollars*), Thailand (*Patong Girl*) and Kenya (*Die weiße Maasai* and *Paradies: Liebe*). The map taken from Belliveau (2016) in Figure 10.1 shows how film settings and sex tourism trends match.

Watching *Sand Dollars*

It starts with images that are common. Tropical scenery, a bit sad. Some actors speak Spanish, for which subtitles are provided. This creates a feeling of authenticity because this is the language of the people. We watch players in the sex tourism business in the Dominican Republic, embodied by Geraldine Chaplin as Anne, the older woman, tourist and client, and Yanet Mojica as the younger one, Noeli, a sex worker. There is also Yeremi, Noeli's lover, who wants to live with her. Noeli is pregnant and Anne is in love (with Noeli). Anne wants both of them to travel to Paris. But Yeremi and Noeli cheat on all their clients and leave them heartbroken.

In a great colonial house, Anne and the other wealthy tourists talk about how everything was grander in the past. This scene crucially shows Anne's diremption by making use of specific language practices and discussion topics. The elegantly dressed White elite, who constitute the new colonizers in the Dominican Republic, spend their evening discussing the economy and the politics of the land, which they claim to know so much more about than the Dominicans themselves. While Anne tries to keep up appearances, the neocolonial discussions fade and Anne's true problem, her love for a Dominican girl, emerges as the focus of the film. The language used in this short scene is a highly elaborated political lexicon which is used to display the great gap between the White elite and the Black working class.

Besides language, gifts create connections between people. A tourist gives Noeli a necklace as a token of his everlasting love. She takes it to the pawnbroker for 400 dollars. Bodies do not have this power: they are delicate, thin, worn out, old and vulnerable. When the pain and the lies become unbearable, Anne phones home. She speaks French, a soliloquy on the phone. Nobody talks to her, except for her grandson, who is telling her about his faeces. Her connection to the world of her descent is trapped in children's language, talk about excrement and silence. French language, however, remains.

Yeremi asks Noeli about what Anne does to or with her. Noeli's gaze is all we see. In this lesbian love story which is also a story of exploitation there are things that remain unspeakable. Bodies collapse, everything is ruinous. Anne meets an old friend. They talk in English. He shows her photographs of a man whom both of them once loved. Now this love has come back. Anne discusses her problem with her friend and tells him that she wants to stay forever. But Noeli does not phone. Problematically, Anne's decision about staying is not rooted in herself but rather in Noeli's behaviour. In a short dialogue with her friend, Spanish is made into the language of emotional blackmail. Spanish, Noeli's language, the language of the suppressed, is used to linguistically emphasize the situation of suppression in which Noeli finds herself. The friend tries to win Anne back to the life they once had together by using French, the language in which Anne converses with her son and grandson.

Anne: But, you know, if she doesn't phone, *na voi* ...
Friend: Well that's good news
Anne: *Me marcho*
Friend: Good idea, you need to leave. *Changer les idées.*

Sad White gaze, wounds that do not heal. The exploitation of southern bodies. Sadness in a luxurious home. American English articulates ancient rights. That the past was better is a topic that needs French. A business call needs English. Personal matters too: 'I'm very much in love with a – Dominican – woman.' And life goes on.

After the necklace, Noeli receives a visa in her passport. Another token of everlasting love. But she will not wear red shoes on the Champs Elysees. She leaves with Yeremi on a motorbike. This is the end of the film, and it is a relief: it is good that everything falls into ruins so that Noeli and Yeremi can raise their child away from late capitalism and neocolonialism.

Watching *Patong Girl*

At the beginning, there are various images: linguistic landscapes of the tourist space, *VISIT BRITAIN* on a bus, a hotel decorated with Bavarian flags. This is Thailand, the mother of two sons says. Here they know three genders (citing the travel guide): men, women, ladyboys. The hotel is not the one they booked. This is dark tourism, catastrophic, ruinous, they say. In dialogues between the German family (the tourists) and their hosts (the Thai people), English is used. All the information we need in order to understand Thai culture is provided in English. Family conflict, in turn, is transmitted in German. It would have been refreshing to have it the other way around – to let the German family solve their domestic conflicts in a foreign language and to have them talk to their hosts in German. A young woman explains the Thai language to one of the sons; subtitles, authenticity. Afterwards, we see impressions of everyday life and people who stare at their smartphones. Language – images – foreignness – Farang – Falang. Pretty – ugly, plastic surgery, sex. The morning after: the Thai girl who is a ladyboy has a book on 'Love Letters'. It is used as a working tool for those who earn a living in the tourism-based sex trade. We are reminded of the language guides used by Philippine maids (Lorente, 2017). The German son does not understand; he wonders why his lover cannot have any children. Smoking weed (son), taking pills (lover). Ruination continues. Aged White men in underwear pass their time in bars, where they are treated badly by Thai people. A stranded White man tells a story about a Thai wife who has poisoned her Farang husband. The wife inherited his wealth. 'Everything goes mouldy in this climate.' We have seen this already in *Sand Dollars*, where true love went bad quickly. Sex tourism in a tropical climate is dystopic, because the sex tourist yearns for warmth and forgiveness that cannot be given. Colonialism is presented as a mental condition that imprisons White men: 'I cannot leave. This is my home', says one of them and speaks of Pattaya. The German mother, meanwhile, performs a maternal form of colonial duress. She wants to help – because this is not a life. Sex work in Pattaya is damaging to those who sell their bodies. Once this has been said, life can go on. Lobster and champagne are served. This is, at least, the climax of the Christmas vacation of an average German family in an average setting of the Global South. And then the holiday is over and the son has to say goodbye to his Thai lover. They have sex in the toilet cabin. At the airport, the son spontaneously decides to stay, like other great lovers of the screen before him, like Corinne in *The White Masai*. Why is it always possible in movies to stay? Where are the visa regulations? In this dystopia because love crosses all borders. A fisherman passes by. The son gets himself fresh cash at an ATM machine. The fourth wall is down and we see him typing and selecting certain services as we seem to sit inside the ATM machine: inside the system, complicit, different, unfree. The son is free, however, especially now that he has cash. He hurries up to the bus station in order to travel to the north where his lover now is. They meet on the bus, by chance, and he plans their life together. But then he has to understand that his lover is not a young woman, but a transgender person, and he gets angry, because he does not want to be queer. We understand that gender equality is just a hollow phrase, a promise nobody intends to fulfil. Isn't colonialism also about homophobia and puritan concepts of sexuality? Therefore, the Thai lover is now going home to where the grandmother rules over everything as a matriarch. She supports transgender identity constructions. The German mother, who has been worried about her son's wellbeing and has decided to stay as well, now arrives there too and wants to speak to the grandmother in English. The matriarch-grandmother replies that she herself is a very well educated person, 'My good woman, I speak five languages – Hokkien, Cantonese, Laotian, Central Thai, Isaan – should I still learn another one [English] at my high age of seventy?' This is cool. The mother is disgraced. Everything now turns into the opposite – the neocolonial visitors seem hapless, silly and generally improper, while the postcolonial inhabitants of this northern town have education, status and poise.

At the non-place, the departure lounge of the local airport, the German mother and her son have a terrible argument about racism, paternalistic attitudes and silliness. Ruination is everywhere, and yet there are moments of love and truth. But in the end, these moments belong to the Thais, while the tourists fade out and remain with their miserable phantasies and their loneliness.

In the end, all of the stereotypes about Thai people and their culture that are presented in the film are given back to the German audience in the timespan of just a few seconds when the lover asks the question, 'What do you know about Germany?', leaving the viewers with bitter feelings about what their own country is worth, when the Thai lover answers: 'I know Hitler, the *Oktoberfest, Sauerkraut* and Heidi Klum.'

Watching *The White Masai*

We begin with a beach and a voice that speaks about memories of a holiday. They should have left it that way: a beach upside-down and a voice that babbles along. Why not? But the voice belongs to a woman who is on vacation at a Kenyan beach resort and unfortunately wishes to stay there in order to live with a man. Therefore, we have to leave the beach.

The man is Jacky Ido, who pretends to be a Samburu warrior, and the woman is a revenant of Leni Riefenstahl: tall, blonde, big eyes, although the author of the autobiography is dark haired and average sized. The beach resort is the Africana Sea Lodge in Diani Beach, now closed. It is as it is, and the babbling Leni-woman goes shopping for dope with her dopey fiancé. 'Look, a Maasai!', he shouts, as we see Jacky in his first scene in the film. There he is, the Maasai, in the very typical pose of the Maasai warrior that can be found in each and every travel guide.

The couple have some problems in Mombasa. Chased by three Kenyans who probably want their money, the chased and the chasers run through the crowded streets of Mombasa without being observed by any of the Mombasan people around. Just as if they were extras from another movie, they are not included in the action of the film. It is then that the two Maasai show up in the middle of the food market as the saviours, the phoenixes. The Maasai-Jacky says, '*Jambo*. Any problems?' The fiancé says, 'We are a little lost. Which way to the ferry?' Big eyes. Subtitles throughout, translating English into German. Then there are gazes. The Maasai-Jacky turns to his fellow Maasai and says something that is hard to understand. Now there are no subtitles. We can only guess that this is Maa for 'Let us show them the way to the ferry'. More gazes. Many Africans on the ferry, but they don't talk to the strangers, again, as if they were not there. Just as they don't seem to be there in Likoni, where the couple are again overburdened by their environment and don't ask anybody for the *matatu* to Ukunda. The Maasai warriors return to the plot as the saviours and guide the couple to the *matatu*.

Woman, fiancé and the Maasai take the *matatu* from Likoni to Ukunda, and are dropped off right in front of their hotel at Diani Beach. This is ridiculous, because in reality no *matatus* go to Diani Beach. The woman says, 'Wanna come with us for a drink?', and Jacky asks his co-Maasai something that is hard to understand. The resort photographer says, 'Picture please. Smile please. Thank you. *Karibuni*.' The resort security guard says, 'Stop! Residents only.' The woman says, '*Sie sind meine Gäste*! They are my guests!' – 'Sorry, residents only.' – 'Wait, wait.' – 'It's ok, [...]' (something that is hard to understand).

While one wonders why somebody should take the Likoni ferry forth and back, wearing Maasai folklore working gear in a Mombasa market, the film goes on. They go to a bar, which they say is the *Bush Baby Disco* even though it looks like the *Shakatak*, a disco that really exists in Diani. 'Do you wanna dance?', the woman asks. Jacky replies something that is hard to understand. The DJ plays E-Sir featuring Brenda, *Moss Moss*: *slowly, slowly we will get there*.

The heat of the night and the music cast a spell on the woman and the others. The dopey fiancé appears and starts a fight with the Maasai. He pushes the Maasai away from his woman and says something that is hard to understand. We assume that this is Maa for 'Hands off'. But too late.

The woman decides to leave her fiancé and stay in Kenya. Had she decided to stay in Thailand instead, she could have seen how the pattern of her film also existed elsewhere, as a repetition of colonial tropes of love and greed and passion, a globalized plot. Just like the take of the fiancé in bed, mosquito net around him. Will we not see this yet again in *Paradies: Liebe*? But it is as it is. She goes back to the *Bush Baby Shakatak Disco* and meets people whom she can ask about the whereabouts of the Maasai. He went home, they tell her.

'Where is home?' – 'Barsaloi, in Samburuland.' – 'How do I get there?' – 'Take a bus to Nairobi.' – 'Nairobi.' – 'In Nairobi, you ask a bus to Maralal.' – 'Maralal.' – 'In Maralal, you ask for Elisabeth.' – 'Elisabeth.' Kenya is a really smooth country to travel in. It seems as if nothing can hinder a tourist from getting around with the only three landmarks that are necessary to find one in a million. Nairobi, Maralal, Elisabeth. The prospect of ending up in a far-away place needs the help of the last landmark, a European lady who can help to find the intended goal: Barsaloi and the Maasai, who is actually a Samburu.

So she leaves: the beach, the resort, her life, everything. And finds the bus and rides on it all night. Next to her sits a woman who is sick. 'Are you ok?' – 'Malaria. Very bad.' – 'Oh.' – 'Are you a doctor?' – 'No.' The stereotype jumps out of this scene and grabs our throats and we are depressed, just as we were when we were connected to Hitler, the *Oktoberfest*, *Sauerkraut* and Heidi Klum.

Maralal is not mentioned in the guidebook. It doesn't matter. It seems as if the woman has been travelling without a travel guide anyway. And without a visa. Her flight has left, her fiancé as well, and the first part of the film leaves an aftertaste of sex tourism that should not have been. It should have been the feeling of wildness, love and liberty. This feeling is transmitted through gazes out of the bus window, where we again find the already familiar pose of the Maasai warrior. We switch off the TV set and go out for lunch.

Watching *Paradise Love*

While Teresa is on her way to a well-deserved holiday in Kenya, everything is prepared for her. Men clean the pool and a bus waits for her and the other newly arrived guests. A tour guide introduces them to the local language, Swahili: '*Jambo! Hakuna matata.*' As Teresa repeats these magic words, the bus reaches the gate of the Flamingo Beach Resort at Shanzu. Staff members sing the song *Jambo Bwana* for the new guests. Long, symmetrical, static takes give it a disastrous ambiance. Then the room, the view, a monkey on the balcony: this is paradise.

Love comes next, as Teresa takes a drink at the bar. She sits with a friend. 'This is crazy. The air. One feels different. The smell of it all.' Her friend has been there before, she knows how it is: these black bodies, how they smell of coconut, how one wants to bite this skin. The biting of black coconut skin can be seen throughout the film: biting of skin while sitting on a motorcycle, biting while dancing. Beautiful bodies with large genitals. She has taught her Kenyan lover some German: *geiler Bock* 'horny ram'. Shrill laughter. Foreignness is tempting and it is quite often flavoured with the teaching of completely unspeakable words, which are not unspeakable due to their taboo meaning, but rather to the unbelievable foreign combination of consonant clusters without any vowel to fix the linguistic problem. Austrian words like *Speckschwartl* 'bacon rind' or *Blunzengröstl* 'roasted black pudding' contradict Swahili phonology. These words are used to expose the Other, in this case Josphath the waiter, who gets ridiculed. More shrill laughter. And Josphath is unable to pronounce the words. We are unable to bear the shrill laughter.

Later, they talk about their pubic hair. They will not shave themselves because here, in Africa, anything wild and natural will be appreciated. 'They take you as you are.' And the friend leaves, with Musa, her lover, on the motorbike she gave him. Teresa goes to the beach and meets Beach Boys. *'Karibu, komm here.'* 'Welcome, come here.' *'Willkommen Afrika.'* 'Welcome to Africa'. *'Hakuna matata.'* 'No problem'.

Later, dancing in a small café and an attempt to have sex. As long as no language is involved (dancing, gazing), the paradise is intact. Once language comes in (at the bar, during sex), it is spoilt. 'Let me give you my gift.' Teresa meets another man, Munga, whom she provides with a sentimental education. But even then, everything is ruinous, love and money, this cannot work. Teresa goes to a small bar and meets another Beach Boy. He is concerned: *Ich kümmer mich an* 'I care on you'. She replies, *'Ich kümmer mich an* – what is this supposed to mean? You better speak proper German.'

Stereotypes and clichés are consciously transferred. In probably the most cringey scene, when a hired male sex worker dances for the four Austrian women, the scene is made even more cringey through its verbal exaggeration of spoken clichés. There are scenes in which the women yell, 'We want an African dance!', 'This is Africa! This is being wild! Is this foreplay?', 'Look, like a carnivore!' and 'Mombasa express!'.

Sex, lies, sadness, greed. A display of colonial cruelty, bodies possessed by others, the colour of the skin as the dividing principle.

Rudolph Valentino's last words are said to have been, 'Don't worry chief, it will be alright.'

Notes

(1) All information about awards is taken from Wikipedia (accessed 12 August 2020).
(2) See https://www.nytimes.com/2015/11/06/movies/review-sand-dollars-unrequited-love-in-the-dominican-republic.html?referrer=google_kp (accessed 12 August 2020).

(3) See https://variety.com/2014/film/festivals/film-review-sand-dollars-1201331316/ (accessed 12 August 2020).

(4) Translation by authors; the original German text reads: 'Literarische Werke, die sich solchen Liebesabenteuern widmen, profitieren von einer westlichen Sehnsucht nach Exotik, der Lust auf Verbotenes, unbewussten Machtinstinkten und/oder dem Bedürfnis nach Kompensation für Ohnmacht. Interessanterweise klappt die Aktivierung dieses Gefühlsmusters nur in einer Konstellation von Geschlecht und Rasse, dann nämlich, wenn die Frau weiß ist der Mann aber nicht. Die Suche nach der Liebe im Fremden und der Rausch in und durch die Exotik werden grundsätzlich durch Frauen verkörpert, während europäische Männer sich als Eroberer, Forscher und Händler exponieren.' See http://journal-ethnologie.de/_Medien/Medien_2007/Das_Bild_der_Maasai_aus_der_Sicht_einer_weissen_Maasai/index.html (accessed 12 August 2020).

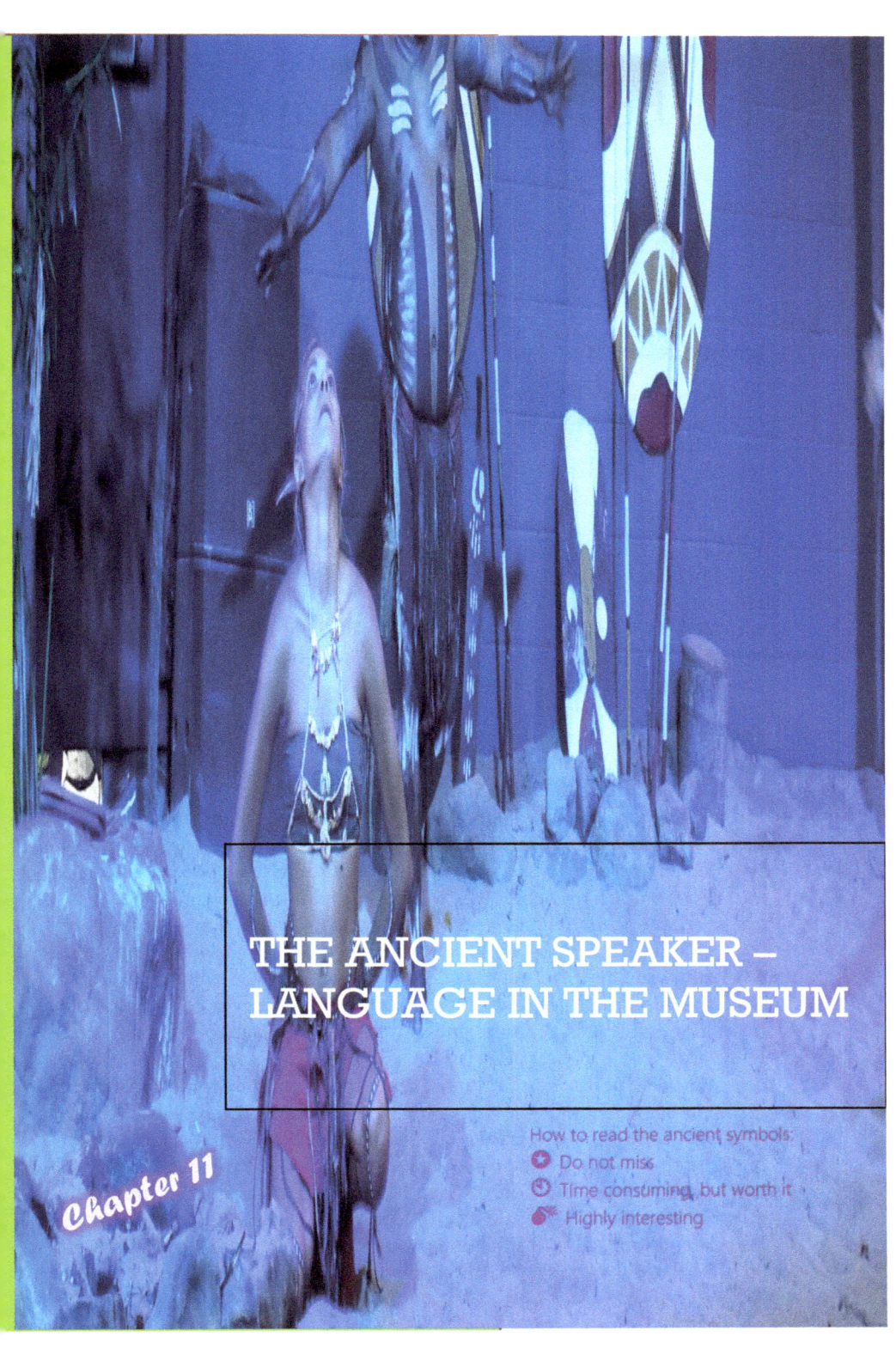

11 The Ancient Speaker

Language and Heritage

Museums serve to preserve. They display, transfer, construct, bore and amuse. And each exhibit is accompanied by language, whether in vast elucidation, in the simple name written underneath an object, or in the words of the guide, directing the listeners through a subjective selection of things. Language is the means to convey the artist's intention to those who do not know and are eager to learn. And often the language *is* the exhibit, used to transmit ideologies and perform the past. Language is often particular there, spoken by a spectral figure that is produced when language is made into an object: the ancient speaker. As a point of departure to the ancient speaker and what this figure might get to say, we want to think about the ways in which language, especially language as an absolute object that can be archived, shown and owned, is constructed as heritage in a linguistically complex postcolonial world. We want to think about the connectedness of objects and performance, and the imagination of the past as part of an arena of binary oppositions, in a setting that is shaped by complex entanglements of colonial, metropolitan, local, Indigenous and global epistemes and practices. By thinking about language as a historical construct and a faint memory of an equally constructed past, we will also have to think about linguistic knowledges, ours and those shared by people elsewhere, and about the terminologies that are available to us when we talk about these issues, and that regularly seem to fail us. LANGUAGE, HISTORY, the PAST and HERITAGE mean different things to the different people we talked to, even though in all the encounters we had, these diverse meanings could never be disentangled from more global, hegemonic ones: language is, in spite of its many available conceptualizations, always also a dot on a language map, a delimited object, and structure instead of practice; heritage tends to be constructed in line with various ideologies connected to hegemonic institutions and practices, sometimes in efforts of strategic essentialism, and sometimes because this might be the only way it makes sense at all as a topic of a discussion or an exhibition. What, then, are the meanings of sites of linguistic memory making and exhibitions of language as part of heritage, prepared or offered by Indigenous communities, grassroots activists and community custodians? Do they present alternative views on language and sociolinguistic history, or can they perhaps teach us something about language in a different way? And why would people employ such a contested, colonial institution as a museum in order to present their own points of view about language as heritage?

These questions will certainly not be easy to answer. They require a more critical approach to our methodologies, metaphors and institutions alike. In her book on *Decolonizing Methodologies*, Linda Tuhiwai Smith (2012: 30) points to precisely this need by provocatively asking, 'Is history important for indigenous peoples?' This question needs to be asked, because history, like linguistics and other disciplines in the humanities (and elsewhere) largely exists, Tuhiwai Smith argues, as an academic field and episteme of the North, where other ways of thinking about time, language and so on are excluded:

> The negation of indigenous views of history was a critical part of asserting colonial ideology, partly because they challenged and resisted the mission of colonization. Indigenous people have also mounted a critique of the way history is told from the perspective of the colonizers. (Tuhiwai Smith, 2012: 31)

We could say almost the same about linguistics here, which shares with history a number of characteristics in the sense that linguistics, too, to a very large extent excludes Indigenous theories of language, for example concerning transcendence, spirituality, heteroglossic practices, embodiment. Instead, academic Northern humanities such as history tend to include, according to Tuhiwai Smith (2012: 31):

- the idea that history [and linguistics] is a totalizing discourse
- [...] there is a universal history [and linguistic structure]
- [...] is one large chronology [and pattern of change]
- [...] history is about development [and language about standardization, literacy, etc.]
- [...] history [and language] is about a self-actualizing human subject
- [...] can be told in one coherent narrative
- [...] history [or, linguistics] as a discipline is innocent
- [...] is constructed around binary categories
- [...] is patriarchal. (Tuhiwai Smith, 2012: 31)

These key ideas and criticisms have been brought into various debates, including in linguistics, but seem strangely absent in metalinguistic discourse surrounding language in heritage contexts. As a consequence of thinking about our epistemologies, an interesting change of perspective takes place: whatever institutions, archives and museums may be about, they are in fact heritage themselves, but an ambiguous one. They are places that are claimed to represent the decidedly local, yet they put forth particular genres of presenting and narrating language (and history) that relate to epistemes of linguistics that continue to reflect the colonial origins of the discipline. Local knowledges about languages have little role to play here, and language as a local practice is enregistered through its inclusion into these hegemonic genres and institutions in such a way that it turns into an artefact. This of course is not surprising, and as argued by scholars in Critical Heritage Studies such as Laurajane Smith (Smith, 2006; Smith & Akagawa, 2009, 2019), any shared immaterial inheritance or experience becomes *material* upon its construction as HERITAGE, not just the immaterial in postcolonial contexts or that which is on display in museums at tourist sites. Furthermore, the type of artefact that language becomes through its materialization is most likely writing – a text, language printed on a surface and resembling language in literacy practice, which is most likely literacy in colonial contexts: normative, debilitating, regulating, rather than poetic and individual. In all its similarity to text, language in the museum and as an artefact has not been made part of discussions about ownership and restitution, almost as if it had been overlooked. In archives and exhibitions, language continues to affirm colonial language ideologies and ideas about language as something that can be made, owned, placed on maps and shown in showcases. No paternalistic gestures of stingy generosity by those who manage the large collections in the North's national museums are directed towards language; what is given back to the owners in attempts to make sense of all the things in the boxes and cartons that have been in the collection for a long time but have not been given any space in the exhibitions are different items. Masks, a whip recently, skeletons and a Bible. Things that can barely be looked at without evoking a sense of disdain: are these artefacts, after all, not the emblems of a colonial violence that refuses to be over? An act of compensation that might not even be felt by those who visit these museums now? Very often, restitution appears not to be a painful act to those who return objects and evokes the impression that museums and their metropolitan audiences are getting rid – of guilt, responsibility, things. What is gained is that the change of the design of exhibitions in these museums might even make them more interesting to audiences than they were before, when people stopped going there.

Language, in the meantime, trickles through the cracks in boxes and shelves. It remains firmly in place, as words on maps and plates and labels, in explanatory catalogues and on walls of exhibition halls that nobody will ever read comfortably. The silences that surround all these words remain as well. In her book on the search for Livingstone, Marlene NourbeSe Philip writes on a Museum of Silence: 'You must return these silences to their owners. Without their silence, these people are less than whole' (Philip, 1991: 57).

In the Exhibition: Language as Dead Matter

There are other museums, we thought: small, community-run museums and exhibitions about local history, in various parts of the colonized world. Such exhibitions and museums, which are part of grassroots initiatives and projects, have only relatively recently attracted increased interest from sociolinguists and others (Carman, 2002; Cuno, 2012), but they now seem to contribute to a greater visibility of sociocultural and linguistic diversity, as well as of a history of social injustice that now can be critically dealt with (Piller, 2016). Sometimes, such institutions emerge out of cooperations with linguists and anthropologists, such as in language documentation project contexts. In other cases, community-run and based museums are encouraged by experiences with tourist business and UNESCO bodies. Christina Kreps, writing on Indigenous curation in Indonesia and North America, observes that an increased interest in ethical negotiations (concerning anthropological collections in museums and archives) was another motivation to create local institutions for the dissemination of Indigenous knowledge (Kreps, 2009: 198). The idea that they all reflect diversity seems to make sense, as these places are concerned with a large variety of different postcolonial contexts and experiences: migrant labour, relocation, resistance and slavery, language loss and cultural alienation, women and work. Such sites almost always offer information on language, as part of the heritage concepts of the community or place, as a means to endow the site with authenticity (in whatever sense), or to express an interest in overcoming inequalities in knowledge transfer. Interestingly, however, instead of celebrating language diversity as part of living and vibrant cultural practice – like music – language tends to be framed as enigmatic and gone in the greater part of the exhibitions on display in the advent of a fundamental and radical rearrangement of the more wealthy and commodified museums. Ways of speaking, represented, for example, by emblematic lexical material written on panels or on colourful patches on maps, were dead matter: they were exhibited in the form of lists, books and manuscripts that were claimed to be impossible to read for contemporary community members, as words attached to old objects or as performance of a way of speaking not known any longer outside the museum.

In Simon's Town (South Africa), the *Noorul Islam Heritage Museum* (Figure 11.1) has been established in an old house formerly owned by the Amlay family, members of the relatively large Cape Malay community who lived in Simon's Town until their relocation during the Apartheid era (in 1975). Even though it was not possible for the majority of community members to get back their former properties in Simon's Town (now a popular tourist destination for whale sighting and penguin watching), Cape Malay people hailing from this place continued to express their strong interests in reclaiming their heritage. In the museum, a large collection of artefacts commemorating community life was displayed (Davis, 2015). In particular, the exhibition contained illustrations of practice-as-heritage, such as meals and language. There was a large table with a white tablecloth and elegant china on it, and on the plates and in the bowls there were

Figure 11.1 Signboard in Simon's Town, South Africa

real or perhaps imitation sweets and candy. Silver cutlery, crystal glassware. On the chairs around the table sat mannequins clad in Cape Malay costumes. A delicate light blue veil hid the face of the dummy of a woman. This was a reconstruction of a wedding feast. Luxury and abundance, communality and intact family life. A very emotive exhibition if one considers the possibility that this was a way of being together that has not been possible since eviction from the site took place. Sedick and Zainub Davidson, the founders and owners of the museum, provided tours through the exhibition. Engaging with his guests, Sedick mostly spoke about what was lost: property, future options, community and dignity: 'we were beautiful people, such beautiful people'. Language fell into the category of things lost as well; it is a particularly important part of the museum, most obviously because the first generations of Malay people to have lived in South Africa played such a crucial role in the establishment of the language practices shared by citizens today. Having been brought to the Cape as enslaved people from south-east Asia by the VOC, they were deprived of their languages very early in South African colonial history, even though, as Muslims, many people maintained knowledge of Arabic. A Bahasa manuscript written in Ajami is presented as enigmatic and impossible to decode: 'This is in our old language. Nobody can read this text today, nobody, not even the people from the university.' The tour continued, repeatedly emphasizing the loss of the language whose material traces are still there – in Ajami writing and artefacts brought from contemporary Malaysia. When asked about simply inviting a person proficient in Malay to visit and translate the document, Sedick replied that this had been attempted, but the knowledgeable people from Cape Town University would not come to Simon's Town, at least not to the museum.

This was a place where tales about rejection and marginalization were sung, putting into rhyme the loss of the languages that were once part of the community's social and cultural life. With the removal of these languages, the monolingual Other emerges: the distant past of the Cape Malay community of Simon's Town is a past where language is pure, a single code owned by a single group.

Yet, even though a single language like this appears to make a past that is otherwise messy, complex and confusing into something controllable and manageable, it is not a language that can or should be used or practised. It is dead matter that can be seen and touched in the museum, but that cannot be read or understood; it is meaningful, though, as it stands for broken links to where there was once a home, for interrupted relationships with other people and for lost dreams and hopes. Language as the heritage of marginalized or diaspora communities here and elsewhere is transcendental in an unsettling way – it is entirely about the interiority of past generations, like spirit language heard during a trance when the possessed is mounted by the ancestral gods. In the community museum of Accompong, a Jamaican Maroon village, examples of Twi are placed next to a selection of photographs taken during Kromanti rituals, while in Seaford Town, a Jamaican German migrant village, wordlists of German food terms were placed next to haunting pictures of emaciated ancestors in the now closed exhibition of the village's history. Again and again, language is turned into an exhibit simply because it has been lost, is dead and cannot be revived. At the same time, this is always only *one* language, never anything that would be part of diverse repertoires or fluid practice. It is named, solitary and was once owned by just one single group.

The same aspect can be seen in a different guise in Cairns, Australia. The Tjapukai Cultural Centre, which advertises with the slogan 'Discover the world's oldest living culture', plays with language across all stations of this 'hands on museum'. In contrast to the small heritage museum in Simon's Town, the Tjapukai museum was designed and constructed for the purpose of being a museum and attracting the tourists who come to this place mainly because of the Great Barrier Reef. Unlike the museums in South Africa and Jamaica, Tjapukai was not initiated by the community itself but by North American theatre and amusement park professionals; yet it is the only heritage museum devoted to local aboriginal culture (and language) in its region and seems to exist largely unchallenged and uncritiqued.

In the museum, besides being introduced to the skills of boomerang throwing, spear throwing and food preparation, the visitors are pelted with fragments of language, to make them feel close to the history and traditions of Aboriginal culture. Although, according to R.M.W. Dixon (2015), the Tjapukai language is no longer spoken, its words play an important role in the performances of the Tjapukai museum guides.

While performing the Aboriginal culture in impressive light-and-dance shows, the accompanying English explanations are constantly interrupted by the Tjapukai translations of the nouns just mentioned, thrown in by the other performers.

Even though the objectification of language and its dislocation in global capitalism are the two fundamental principles that form the language in tourism (Jaworski & Thurlow, 2010), the obtrusive and inflationary presentation of Tjapukai words during the dance performances may be seen as a form of corporeal politics (Henry, 2000) in order to gain agency, revitalize identity and participate in representations of cultural authenticity. The young guides at the museum, who are members of contemporary Australian society, transform themselves into actors in an imaginary ancestral culture and language and use 'broken English' to explain the past. When we happened to talk to one of the guides outside the museum, the 'broken English' was gone and the young woman turned out to be a graduate with a degree in performing arts, speaking Australian English as most of the people around there do. Neither she nor her colleagues really came from Northern Queensland; they were professional entertainers who had studied Dixon's and other linguists' work in order to be somewhat familiar with the language that once existed there. In their performances, language was exhibited almost as a wordlist of the most important nouns that should be remembered and stored for posterity.

The Ancestors: Inverted Others, Tidy Pasts

The construction of the speaker as a revenant from the colonial past and the performance of language as a wordlist do not just happen. They are metaphorical and performative parts of language ideologies that have been spread across the globe: epistemic violence clad in a coat of heritage and education practices, now, it seems, embraced by all those concerned. The Northern and metropolitan concepts of language and the ideologies pertaining to the monolingual mother-tongue speaker (Bonfiglio, 2010; Gramling, 2016) here appear as the only possible way to conceptualize language and talk, and they are used as a means to construct the (dead, ancient) speaker of a (lost, ancient) language as the ideal monolingual: where linguistic diversity and the messiness of everyday talk are seen as problematic and bad, the ancient speaker suggests that once upon a time, perhaps in precolonial times, order, monolingualism and clarity also existed among those who are marginalized and seem to be left behind today.

Discourse in heritage tourism therefore focuses on the colonial context as an experience of harmful mixing and blurring, and on creolization as a process of interruption and decay. This discourse presents literacies-as-traditions as eroded through diversity, and language-as-tradition as dead matter. With the ancient speaker's language being gone, the ancient speaker himself (it is always a man, it seems) turns into some kind of inverse Other – the Other as super-Northern, hyper-monolingual, purist and traditional. One's own past and ancestors are constructed in line with hegemonic Northern imaginations – before ruination, this all was like Europe, these exhibitions seem to tell us. These representations of language repertoires are about something deeply rooted in the past, about individual languages as obscure and enigmatic remnants of an imagined, formerly more complete, more orderly world.

Yet these museums are about a very painful experience: life in a complex and linguistically unjust reality, where the binary contrast between an imagined tidy past and a complex present is supposed to evoke feelings of perdition and loss. Exhibitions mourning this are not something out of which pours the 'Indigenous voice', or that shows other epistemic possibilities (Southern theory, Indigenous epistemologies ...), but something that illustrates how hegemonic and powerful metropolitan and colonial thinking as well as the practices surrounding cultural politics are. What makes these places worth visiting and attractive is that being there makes you feel quite good: as a researcher, activist, critical thinker, you do good to those who have next to nothing, and you express a reflexive and concerned attitude towards coloniality and imperialism. Here, heritaging seems to produce ultimate appropriation.

What has been presented here as insights into a project on the banality of commercialized educational trips and globalized tourism suggests that banal, tourist-oriented practices at heritage and cultural tourism sites include performances of lost language, by projecting language maps onto a floor where dancers move. It is not only language that is projected here, but also images of people: the monolingual Other and his audiences, being required to mourn and utter some words in response to each other.

What makes such banal settings important? The heritage museums in which language is now shown are relatively new initiatives themselves, but – and this is crucial – the institutions (museums, archives) are old (Stoler, 2009). They stem from colonial contexts, and as such require certain genres in narrating and performing history and heritage which force upon performers and audiences ideas about language that are totalitarian, monolithic and absolute. The ancient speaker in this genre is a zombie-like creature, a counter-image of the Self – not exposed to messiness and confusion but existing in a setting of clarity and unambiguity.

Outside the Exhibition: Language of the Unlost

Today the town of Binga lies beneath fields of guinea corn and beans. After harvest, its remains can still be seen: a potsherd here and a hut's outline there. With each agricultural cycle it becomes more invisible, and with each season some of the ancient mango and baobab trees at the site are gone.

A perfectly round and polished sphere of iron lay between the heaps of freshly tilled soil. It seemed to have been all that remained of the ironworks that once were plentiful at this site. Picked up, put in a pocket with a hole in it, and lost.

The Hone people who live in the villages around Binga still consider it to be one of their sites, even though nobody lives there any more, and it is not even a place of interest to archaeologists and historians. This part of northern Nigeria is rich in ruined sites, with remnants of old fortification walls and Persian gardens, the debris of mosques and houses and huts. Binga was destroyed during the jihad wars which the Fulbe led against the population of the entire area in the early 19th century, or during the violence that ensued afterwards. Large pots which contained items of spiritual power and which not even the conquerors dared to touch remained *in situ*, under some baobab trees, well into the last days of the British colonial era in Nigeria. At some point in the 1950s they were removed. Many other objects of pre-Islamic origin were taken as well, and the childhood memories of Mohammad Hamma Dada, who died many years ago, were clear enough about where these things went: British colonial administrators had taken them, he remembered, in order to erase pagan history and practice, and in order to put them into the museum for research.

The museum is still there, a large building in the city of Jos, dating back to 1952 and standing next to the reconstructions of old architecture – mosques, villages, city gates – as if the ruined towns of the savannah had been resurrected where their artefacts – pots, weapons, iron – were now kept. Mohammad and many of the older generation said that the museum's archive, which they claimed was downstairs in the basement (did the museum have a basement?), contained boxes and boxes of objects that were once taken away from Binga and Pindiga and all the other old settlements of the Jukun, without asking for consent and without proper preparation. A fatal error, some claimed, as many of these items, such as the unspeakable things from out of the big pots that still stood where Bima once was, had spiritual agency that could be dangerous if treated inappropriately. Many of these things had always been secret, and had never been shown to non-initiated men or to the women and children. The objects were part of the spiritual practice of the shrines and sacred forests, and had a connection to ancestral kings and village founders. Now they were in the museum, thrown into cartons in a disrespectful manner. The museum alone, they said, was already too revealing. The building unequivocally suggested that there was something inside, hidden yet visible, and therefore unveiling secrecy and desecrating sacredness. A basement filled with stolen spirits that continue make noises that only the initiated elders were able to control. The language spoken by these objects is a language of remembering conquest and ruination, asking for a place in which to lament.

CHAPTER 12

SPICES AND FLAVOURS
LANGUAGES AND WORDS
DISHES AND CONTINENTS
***
COOKING AND PEELING
EATING AND BUYING
NAMING AND ORDERING

12 Cooking Class

Soup and Halva

Dinner is served immediately after leaving Wadi Halfa; and the sensation is certainly a strange one of feeling oneself carried along by a powerful locomotive across deserts which some eight years ago were virgin, whilst discussing a dinner which would do honour to one of the best modern hotels. As a matter of curiosity, I give the menu:

POTAGE JULIENNE.
POISSON BOUILLI, SAUCE HOLLANDAISE.
GROSSE PIÈCE DE BŒUF GARNI.
PETITS POIS À L'ANGLAISE.
POULET RÔTI.
SALADE DE LAITUE.
CRÊME RENVERSÉE.
DESSERT ET CAFÉ.

(de Guerville, 1906: 242)

***

The grandeur of the Ngorongoro Crater elevates every wedding to the sublime. The transcendent views, thick carpets of rose petals strewn as far as the eye can see, traditionally attired Maasai choirs, a personal butler for every wedding guest, wondrous bridal banquets ... a truly unforgettable start to a charmed honeymoon.

Halva Rosewater Ice-cream

5 eggs
125g caster sugar
750ml cream
250g halva – grated

Whisk the eggs, sugar and rosewater together until very thick and fluffy. In a separate bowl, whisk the cream until soft peaks are formed. Fold in the halva. Fold cream and halva mixture into eggs/sugar mixture. Freeze.
Serves 4

(Short, 2004: 154)

Wadi Halfa itself does not exist any longer. It now lies underneath the waters of Lake Nasser. The villages have been moved elsewhere, having had to make room for national parks, mining, roads and railways.

Everything else remains firmly in place: the opulent dinner, the relationship between technology and luxury, the exquisite remoteness of the dining room in the midst of grand nature. Talk about experiences of elegant ingestions, the culinary rhetoric of staged authenticity, the scribbled recipes from exotic cooking classes: these are the new-old linguistic practices of Orientalist tourism.

Culinary tourism, Melissa A. Baker and Kawon Kim (2019: 252f.) write, is based on notions such as authenticity and heritage. Food provided to clients in tourism settings is considered authentic if it possesses 'the characteristics of a particular region or cultural traditions' (Baker & Kim, 2019: 253), and dining experiences are authentic when they represent 'national cultural characteristics' (Baker & Kim, 2019: 253). Moreover, food tourism is often based on a set of practices that are characteristic of heritaging, such as through highlighting differences and boundaries between social groups, the historicity of specific practices, objects and places, and the uniqueness of 'tradition' (Smith, 2006). Food here is not consumed, in other words, in order to savour well-composed ingredients or in order to be nourished, but in order to experience its context: decoration, architecture, environment and terrain, people, language and time. Therefore, the dining experience in a safari lodge near the Ngorongoro crater in contemporary Tanzania offers authenticity and heritage experiences in spite of its colonial ambiance. The 'personal butler' and excess of luxury, such as the 'thick carpets of rose petals', strikingly resemble the luxury and amenities aboard the colonial trains on which elaborate dinners were served. Colonial practice and colonial constructions not only remain firmly in place, but are reinforced in these places and, more precisely, in food tourism practices at these places (Hall, 1994). The message is clear: Africa has only vaguely existed prior to its colonization, as the 'Maasai choirs' may suggest. The Africa that can truly be experienced, seen, travelled and consumed only becomes accessible through the construction of railway tracks and safari lodges. Before this, there was nothing but 'virgin' deserts and 'grandeur'. And through colonialism, the narrative of the *Kitchen Safari* suggests, Africa emerges – out of an Empire whose militaristic costumes and game reserves now turn into the characteristic features of the postcolony (Mbembe, 2001). Heritage and authenticity are represented, or achieved, through colonial performance, gaze and discourse. Halva rosewater ice cream instead of *crème renversée*, and we don't need to go back in order to have it once more: there is a recipe, which guarantees that the experience of the colonial sublime can be tasted, once more, at home somewhere in the refrigerated Global North.

Many sites of colonial heritage gastronomy remain elite tourist destinations, such as the Raffles Hotel in Singapore as the site of the 'authentic' *Singapore Sling*, while others are constructed in different, more 'contemporary' ways. Yet the story that is told at these sites remains closely connected to colonial experiences and paradigms. For example, the Spice Tour on the island of Zanzibar is as deeply entrenched in colonial power relations that continue to be meaningful as the luxury safari in the national parks on the East African mainland. Yet the Spice Tour feels slightly different; it aims at displays of rawness and purity.

Spice Tours can be booked almost anywhere on the island. Unlike the exquisite lodge by the Ngorongoro Crater, spice plantations are easy to reach and not expensive to visit. Tourists can book guided tours at their hotels and guest houses, at bars and cafés, travel agencies and shops. Even though the dates for their visits are usually arranged according to their individual needs, the tours themselves take place along busy paths. Clients are taken to the various spice farms, which offer guided tours, cooking lessons and culinary experiences that are closely connected to the plantation: smelling, tasting and seeing the spices that form the basis of the curry as much as of the Sunday roast. The tone of the tour is personal and jovial; there is joking, a feeling of spontaneity, a seemingly uncomplicated encounter with the raw and uncooked. Clients are taken in minibuses to the plantations where they are greeted by a guide. The tour starts close to the small parking lot, where there are already some spices – hibiscus, frangipani or perhaps curry leaves. The guide asks about nationalities and cooking habits and takes the group from one plant to the next, peels a bit of bark from a cinnamon bush, puzzles his customers by explaining that pepper is a vine and not a tree, and continuously offers leaves and fruits to smell and taste. At the end, there is a small shop or a market with packets of spices, baskets full of colourful things, and food and drink.

The advertisements for Spice Tours often promise much more than this, such as culinary safaris, food tours and so on. Even though a Spice Tour never takes more than a few hours, including introductions, shopping and a meal, it is offered as a journey, like a safari, as on the website of the Australian agent *Urban Adventures*[1]:

> We begin our Zanzibar food tour at the source – a local spice farm. Here, you'll get a chance to taste seasonal samples and learn local farming techniques. Think Zanzibar only farms spices? Think again! On top of your spicy favourites like cinnamon, cardamom, ginger, nutmeg, chilli, vanilla, saffron, and curry, they also cultivate pineapples, jackfruits, lemons, limes, oranges, pomelos, durians, and coconuts which you may get to taste straight from the tree!
>
> After snacking on a variety of fresh fruits, we'll head over to a nearby farming village. This is a rare opportunity to explore a local village off the tourist route and experience local life, work, and development just as it is.
>
> From there, we'll visit a local farmer to see her small cassava farm. After trying our hand at African farming, she'll take us to her outdoor kitchen for a local cooking lesson. All the ingredients used for the vegetarian dish are fresh from the village and nearby farmland. Ever wanted to know how to make coconut milk from a fresh coconut? Now's your chance!

Our Zanzibar tour will end with a tasty Swahili lunch, including the dish we prepared as well as spiced rice, bananas, and local in-season produce. Feeling satisfied, we'll head back to Stone Town. On our way, make sure to ask your local guide about other off-the-beaten-path adventures in Zanzibar.

[...]

Inclusions: Local English-speaking guide, transportation, lunch, cooking lesson.

Exclusions: All spices and other products sold at the farm, tips and gratuities for the guide.

Dress standard: Modest dress and good walking shoes for the farm.

Off the beaten track, original, authentic, rare. The Spice Tour leads to a deep experience, just like the safari to the Ngorongoro Crater. And like the wildlife tour, it is tightly entangled with the metropole. The plantations have never been places of Indigenous sovereignty, or even remote. They were created, like those on other islands of the Indian Ocean, in order to sustain colonial trade networks. Clove trees, which were the most important plant of the Zanzibari plantation economy until the end of the 20th century, were introduced to the island only in 1818, when the government of Seyyid Said bin Sultan resulted in the incorporation of this part of the Omani empire into global trade networks. The ever-increasing need of emerging European bourgeoisies for spices, tea, coffee and so forth resulted in the drastic growth of plantation economies in the Global South, not only including the introduction of neophytes to islands such as Zanzibar, but also the growth of the slave trade and forced migrations. Right from its beginnings, the spice plantation economy was closely connected to slavery and colonialism, to the creation of the modern nation state and the colonial empire, to Europe's working class and urban poor, to capitalist markets and global exploitation. The plantations are therefore also closely connected with food such as cinnamon rolls, *Currywurst* and vanilla pudding, which have a 'traditional' Northern feel and yet were originally based on the availability of moderately priced spices from the South.

Since the decline of the spice export trade (and the rise of the artificial flavour industry), the plantations on Unguja, the largest island of the Zanzibar archipelago, increasingly depend on tourism. Mahenya and Aslam (2014: 99) give figures of up to 99% of the plantations operating as demonstration sites and not as working farms any longer. They remain lively places, where authenticity and heritage are explored in particular ways. Groups of tourists and their guides walk there, climb, peel, taste, try, cook. There is often a feeling of realness, of accomplished self-actualization, of authenticity. In her analysis of the Spice Tour as part of the growing global ecotourism industry, Honey (1999) describes the transformation of the plantations into places that offer precisely this.

Colonial erasure, feelgood epistemes. A young man climbs a coconut palm tree and demonstrates plantation work as plantation play. Sitting there on top, he sings *jambo / jambo Bwana / habari gani / hakuna matata*. A group of tourists take out their cameras and smartphones and video his performance. The guides have a break. Laughter. Comments and questions. Languages overlap, mix into one another, blend. Another young man introduces himself as *007*, the *James Bond of Zanzibar* and walks away, shrugging his shoulders. Why do we know his name now that he has left? But then, why should we know all the things we now know, about how pepper is grown and cinnamon is harvested? Such authenticity will only remain meaningful as an anecdote, a story to tell at home, later. What if it is not simply plants and fruit that have been changed in their meaning, from trade goods into non-objects that have but one purpose, namely to amuse? What kind of thing is language turned into in the authenticity setting?

By the end of the Spice Tour, we know that cloves are called *karafuu* in Swahili and that pepper is *pilipili*. Nutmeg, *kungumanga*. Cinnamon, *dalasini*. Vanilla, *vanila*. Strawberry, *stroberi*. At the souvenir shop, there are little baskets filled with packed spices, which make authentic souvenirs. A bag made of palm tree leaves bears an inscription: *Jambo Zanzibar*. The guide who has explained all this to us, politely and patiently, says that he has learned something as well. A few words that might come in handy with the next group of tourists. *Erdbeere, Tasche, Andenken*. Strawberry, bag, souvenir.

Pepper Journey

The Spice Tour, like the botanical garden, is about essentializing space, associating plants with continents and vice versa – Zanzibar, this hybrid space, is Spice Island, a place where diversity is grown – and with food that is turned into national food. Communality and encounters thus gain contradictory meanings as they are offered as concepts of homogeneous community, where the tourist may be present as a guest. That food and feast are occasions of inclusion, of hospitable encounters and togetherness is not at stake: this is about nations eating national food, made from national trees. Ethnic communities eating ethnic food, made from ethnic ingredients. Exotic locals, gazed at by their clients.

Language is conceptualized and distributed equally. The term 'multilingual' makes sense here, as we are constantly reminded of the presence of different languages. One is from Norway and speaks Norwegian, from Germany and speaks German, from Africa and speaks African. Swahili as the language of Africa, Tanzania and Zanzibar erases all other languages present, such as the many languages from the mainland that are also used there; it makes diversity easy to handle. *Hakuna matata, you already know some words*. Like the graffiti in Christiania, Swahili represents a continent's hospitality, as a phantasy of managed hearts and tongues.

The compartmentalized performances of places, spices, heritage and history are multilingual performances by both guides and guests. Yet their fluid and creative language practices are usually not seen as artful and reflected, but as a necessity and as an aspect of commodification (Heller *et al.*, 2014b; Schneider, 2016, among others). The tourist guide Bwana A. describes his personal repertoire of languages needed for his work as follows:

> I'm speaking almost six to seven languages. I speak Swedish, Norwegian, Dansk, Italian, French, a little bit – a little bit of Chinese und Deutsch a little bit. Chinese I learn from my brother who lived there for fourteen years; I learn from him. The other languages I learn mostly from the tourists during the spice tours. Maybe I'm a quick learner, I think.

Speaking and separating many languages here means being able to use many different and separate codes for a specific purpose, namely to work in a tourist setting. In terms of advertising his communicative skills as a guide, Bwana A. never mentions the other languages he speaks – Kiswahili, Arabic and English – nor the semiotically complex performance that is part of his work as a guide. The former he considers irrelevant for a description of his multilingual repertoire, because they are just there, as a result of one's socialization and education, and the latter because performance is not part of the hegemonic discourse: LANGUAGE is words and phrases, not gestures, song and movement. Languages are, in other words, named and counted and exhibited as commodities that are controlled by individual players when they are used in contexts where very particular language ideologies prevail: concepts introduced in colonial contexts, framing language as bounded object and tied to ethnicity or nationhood. Spices likewise are trees and shrubs; they can be combined in order to be sold as mixes, but under specific names: *butter chicken mix, biryani mix, masala spice, coconut curry mix, chai masala.*

This also applies to food itself. In a restaurant by the sea, just in front of a number of historical buildings in Stone Town, the menu provides us with the order of things. Before one gets to the dishes on offer, there is a chronological outline of Zanzibar's culinary history. Food as heritage and living history, as in the plantation. Pepper grows on trees, as a vine that climbs on them. Pepper is *pilipili*. Pepper is a motif, a trope. But how does it get here? Pepper is a journey.

The menu has a heading: 'Zanzibar rich cuisine influences'. White letters on turquoise-coloured wood (see Figure 12.1). In the land where pepper grows, the 'First Habitant of Zanzibar were Bantus from Tanganyika Fishermen.' We need to know this as much as we need to know everything else. 'Choose your starter from their history.' Everything is in place already. Explore. Find out. Yet what, precisely? Perhaps the many meanings of sadness.

Figure 12.1 Chronotopes on a menu

Isn't this, too, worth a closer look? 'Bantu' as an ethnic term is no older than colonialism; the word was coined by Wilhelm Bleek in 1862, when he wrote about typological differences between the languages of southern Africa. *Bantu, Bushman,* terms invented in order to make order. Back to the menu. Bantu fishermen came and created a starter. They came from *Tanganyika*, already a blend, a linguistic masala, consisting of *Nyika* and *Tanga,* which are not just any toponyms, but place names that are placed in a contested colonial space. Later, after the revolution, this toponym blend was to be mixed with another one, *Zanzibar,* in order to create *Tanzania.*

On the menu, history moves on, a linear process from starter to dessert. '9th Century Omani, Yemenis & Persians began their influence and brought along with them new dishes and ingredients.' No political economy of trade, no imperial formations of the Indian Ocean, no plantation and exploitation, but influence and ingredients. In case we are worried about too much pepper in their dishes, we can move on: 'In the 15th & 16th Centuries Portuguese began travelling to the African great lakes including Zanzibar.' Their culinary legacies are on offer. There is some deep logic in these geographies. Travelling to the 'great lakes including

Zanzibar' in that past refers to travels to places connected through the slave trade; today they are connected through these historical experiences, as well as their erasure through tourism.

A nearby shop sells souvenir mugs, fridge magnets and other items with a Maasai warrior, Kilimanjaro and the inscription *Zanzibar the Land of Hakuna Matata* on them. Soaps and shower gels are on offer as well, and bear a more oriental design. The menu explains: 'In the 17th Century, in 1651, the Omani sultanate influenced Zanzibar and brought stronger relationships between Indian & Zanzibar.' Feelgood epistemes, *Khaleeji* influencers investing in relationships. From African beginnings to these ties to India, this gets more and more elaborate, and increasingly has a modern feel, of *chai masala* and batik shirts. Where will this end? 'Around the 20th century, British, Germans & Europeans finalized the multicultural journey.' As in any narrative that is based on thought that also informed earlier narratives, like the Hamite hypothesis of the 20th century, development ends in Europe, and the culmination of history is White. The colonialists of the last century have merely completed a 'multicultural journey'. How nice! How sad.

'Our Signature dishes were created around these rich multicultural fusions.' Those who had to pay for the fusion do not appear in the picture. They have been named into complete invisibilities, Bantu. Someone says *Jambo, have you made your choice?*

Tourism is colonial practice and characterized by the continuity of imperial formations, Hall and Tucker (2004) have argued. This continuity is reflected in the presences of specific forms of language, discourse and representation. Tourist settings in the postcolony are represented in images, advertisements and tourists' discourse as spaces where encounters with the Other-as-host can take place, and where cultural production and language practice are fundamentally *different*: not pure but mixed, not real but invented, not modern but traditional. Linguistic and cultural differences are exaggerated as part of a place's authenticity and worthiness (e.g. as a heritage site, an allochronic site of Indigeneity, and so forth), as in this menu. The simplicity of the binary oppositions presented is self-explanatory; translators and interpreters will not be required unless the tourist decides to explore – for example on a spice safari – the 'authentic' realities of the hosts. Very much like expeditions during the colonial era, tourists will then enter a kind of *terra incognita*, where the 'exotic' prevails; in line with what Fabian (2000) has observed about the expeditions of the 19th century, which mostly used well-trodden paths, today's touristic ventures out of hotel lobbies and secluded holiday resorts and into 'otherness [which only] makes a destination worthy of consumption' (Hall & Tucker, 2004: 8) take place mainly along pre-existing pathways and through commercialized activities. The menu puts them in order, across time and space, while we make our choices as we lounge by the beach.

The beach ceases to be particular precisely at this point. The same applies to anywhere else where a connection between food and hospitable encounters, language and shared environments, the colonial past and the concept of difference (or should we say diversity?) can be made. Cooking classes, guided tours to markets and visits to community museums all seem to be based on the idea that the postcolonial Other's food and ingredients are interesting to know not because they represent some kind of measured wealth (as in Europe's more distant past), but because they are a key to Otherness and to the spectacular.

Poisonous Stuff

Food tourism is about finding ways of including exotic, Oriental culinary experiences in tourism encounters. Spices transport sensual – and sometimes erotic – experiences, an experience that can always be exceeded. In this sense, the Tjapukai Cultural Park[2] in Cairns, Australia, displays plants, fruits and seeds that were used by Aboriginal people for both medicinal and culinary purposes. Here, the toxicity of the plants is constantly emphasized during the presentation, and the possibility of poisoning others or oneself either deliberately or accidentally is repeatedly mentioned. The toxicity of Aboriginal plant use corresponds strangely with the erotic sensuality of the Zanzibari spices, the intention being to highlight exactly the dangerous aspect even on the website:

> Join the Tjapukai women for an insight into the ancient medicinal and culinary uses of native plants, fruits and seeds, which the Aboriginal people gathered. Learn how toxins are leached from poisonous rainforest plants and interesting facts like how flowers speak to you, telling you when the seasons will change and when it's time to move hunting grounds.[3]

As in every other section of the Cultural Centre, the names of the plants in the Tjapukai language are thrown in during the guides' explanations, leaving the visitor with zero remembrance of any of the terms, but with the sentiment that food preparation is highly dangerous and requires highly elaborated knowledge. It is also made more authentic for the tourists, as the guides at the museum are dressed in traditional costumes with their bodies painted in order to fulfil the promise of what the Cultural Centre stands for: 'The original authentic Australian indigenous cultural experience'.[4] The exotic and mysterious Aborigine is kept alive in order to stimulate and produce the touristic melancholia after empire (see Gilroy, 2005).

The Other's culinary habits such as the sensual spices of Zanzibar or the extreme or toxic foods of Tjapukai are stirring and fascinating and inspire various forms of guided tours – not only through spice markets and cultural parks, but also to other people's own gardens and fields.

Since not all people have the same access to recording technology or the internet, the possibilities of profiting from culinary tourism vary immensely in terms of what can be offered. If the menu consists of, let's say, the five most common plants that are grown in the host's garden, the interest of the tourists can only be raised by the most minimal offer in order to convey the poverty and need of the society, which is as escapist as the toxic plants of the Aborigines. One of these minimalistic food tours was attended in Masindi, Uganda, offered by the Boomu Womens Group, situated at the Kichumbanyobo Gate to Murchison Falls National Park. As the sign (see Figure 12.2) shows, various offers invite tourists to join in. The cooking tour is a guided tour in the nearby banana plantation, where fresh plantains and banana tree leaves are cut and shaped in the presence of the visitor, and brought back to the hut, where the tourist learns to prepare them for cooking (Figure 12.3). The only necessary ingredients are the plantains, the leaves and water. The carefully peeled plantains are wrapped in the leaves and left to cook for the next two hours.

The Boomu women's camp has seized the opportunity of tourists' interest in Indigenous African food and offers the experience of extremes, which are created by the highly different lifestyles of guest and host. Food as the basis for life in any society raises special interest in the culinary touristic discourse, whereby not only the minimalistic ingredients and preparation play a role, but also the aspect of eating in what may be called an Oriental manner, i.e. eating with the hands. This localism in food is often opposed to the culinary cosmopolitanism of the tourist, whereby

Figure 12.2 Time eating away a signboard

Figure 12.3 a, b & c Cutting prior to cooking

cosmopolitanism is mainly used to describe places with a large assortment of 'global foods' (Molz, 2007: 79). Molz (2007: 79) further argues that culinary tourism 'is not necessarily about knowing or experiencing another culture through food but rather about using food to perform a sense of adventure, curiosity, adaptability, and openness to any other culture'. This becomes particularly obvious in the cooking classes at Boomu Women's Camp, as the tourists adapt and socialize, experiencing the lifestyle and the food in their own extreme ways.

Extremes, scary food, toxic food, localist food, erotic spices – these all belong inseparably to the tourist's experience of Otherness in culinary tourism (Long, 2004; Mkono, 2011; Molz, 2007), presented in more or less complex guided tours. They all belong to the ultimate experience or, as Urry (1990) calls it, the tourist gaze.

Notes

(1) See https://plus.hertz.com/things-to-do/O3F8b8Hxx8-zesty-zanzibar (accessed 12 August 2020).
(2) The Tjapukai Cultural Park was closed down as a result of the Covid pandemic in 2020.
(3) See https://www.tjapukai.com.au/aboriginal-bush-foods-and-medicine (accessed 12 August 2020).
(4) See https://www.tjapukai.com.au/ (accessed 12 August 2020).

13 Glossy Glossary

In conclusion, the mosaic of the chapters on emotional and physical experiences of wellbeing and escapism, melancholy and destruction (Chapters 2–5), depicted some contexts of language in liminality – beach, transgression, time travel and transformation. The subsequent parts of the composition dealt with linguistic realities of disrupted landscapes, communities and interactions (Chapters 6–8), and with questions of language as a commodified, hostile, estranged and misleading thing not really shared any longer among people in mass tourism settings (Chapters 9–13).

It is remarkable how many aspects of humanity and conviviality get lost in the process of what we have described: a wealth of hospitableness and encountering which would have been an important part of social and cultural life in the various communities in question. While such observations are largely ignored in many linguistic publications on the languages of the tropics, Global South and the postcolony alike (as if this would make a difference), they are crucial in reflecting the situations in which speakers are often forced to interact.

The state of emergency that is everywhere in these chapters and settings asks for even more disruption, we think, in the ways in which we write our conclusive remarks: a glossary, which is nothing more than a few more colourful mosaic stones thrown onto the table.

Scratch card

On a flight with a low budget airline, scratch cards were sold for the price of 2 Euros apiece. The steward announced the sale of the cards over the microphone and explained that the profit would be used to do good. The airline, which is constantly in the news regarding the low income and bad working conditions of its staff, engages in charity tourism. What is behind it? The scratch card names five humanitarian aid organizations that the airline supports. A small picture shows the handover of a cheque on which one can hardly distinguish the sum that was donated to one of the organizations. The very small picture shows the very small letters saying 'ten thousand Euros'. It is not a large sum, we assume, that is gained through the manipulation of the emotions of travellers, who on their journey are pressured to invest pocket money in charitable trusts.

Similar strategies are used elsewhere, on diverse flights and with diverse interests: data-extraction, bonding, rehabilitating.

Taxi

The taxi in Cairo is a haven, even a home. It features fragrance, soft cushions and even a spiritual object – the rosary hanging from the rear mirror. All objects help to build common ground, the basis of communication between the driver and the client. In the glove box there is a guest book, a photo album and a collection of letters of recommendation.

Pyramid

As the epitome of the tourist attraction, the pyramid is everywhere. In the ideal case, it serves as a screen for sound and light shows.

Bingo

In the holiday resort, bingo enjoys a somewhat ambiguous reputation. On the one hand, it is a game that is associated with the monotony of all-inclusive consumerism, and on the other hand, it enjoys a kind of cult status, given the splendid things that one can win. For example, a black and golden glitter notebook for future expeditions, or a purple cocktail shaker for blue hours. For linguists, bingo is not only a treasure trove of paraphrases and metaphors, but also a way to obtain better things and lives.

White paint

Wood painted white indicates organic food, fish and salad. Aquamarine accessories indicate a breeze on a hot day.

Delay

The delay of a plane or train or guide is terrible. But it is necessary: 'That what travel means is not just misfortune but seeking misfortune. In some sense, wanting it. And it is of such crucial importance to us that we must call it religious', as Frederick Ruf (2007: 4) writes in *Bewildered Travel*.

Forgetting

One always forgets vocabulary once learned and then this is the moment when one thinks that one knows nothing. Augé (2001), however, states that forgetting is elemental for society as well as for the individual. One needs to know how to forget in order to be able to appreciate and seize the present moment.

Exotic African liquid

Cream liqueur is difficult to advertise, and therefore it is given remarkable names. *Amarula*, for example, and *Kahlua*. A drink, but the images and songs in which it features seem to have a stronger impact than the sweet alcoholic beverage itself. The bottles evoke emotions and imply a desire for something wild, free. *Amarula* is safari. As a song, it is sex. The name is pronounced as rhythm. Not word, but rhythm. And it bears in itself exoticism and happiness. Like the medicament *Umckaloabo*, which cures nausea or flu, the effect of this liquid is prepared by the rhythmic chorus through which it is advertised. In this way, these concoctions resonate imagined 'age-old' knowledge of 'primordial' societies to heal and make things beautiful.

Culinary entitlement

Our host says: 'The more money is involved, the more trouble you get.' In the upscale resorts where upscale tourists stay, the promised free sundowner that someone might have forgotten to serve (a waiter, a waitress) is a legal affair. The buffet that offers a selection of 10 different starters and 11 mains is a site of gluttony where clients find it hard to miss out on even a single item. What has been paid for has to be made available, has to be ingested. The notion of entitlement in tourism wherever its protagonists have the verbal and material power to claim whatever they think is theirs creates precisely the environmental destruction and consumerist outfall they so despise. Entitlement to water results in a poisonous gift of small plastic bottles, which will not decompose within the next 800 years.

Collectibles

Plane-shaped salt dispenser. Foldable hairbrush. Foldable toothbrush. Sleeping mask. Cake of frangipani soap. All-inclusive ribbon. Ballpen and notepad. Beach towel. Stickers. Pebbles, sand, seashells. Phone numbers. Memories.

Barter trade

Things for feelings. Prior to the trip to the resort in Africa, things that are no longer wanted are put into the suitcase: clothes, pens, shoes, a watch. At the end of the holiday all this will have been given away, donated, gifted. After check-out, a T-shirt and a pair of flip-flops remain in the room. Gestures of gratefulness and smiles are the reward. Letting go of the things that make life tedious is grounding. Junk export works where Othering takes place, where those who receive leftovers are considered sufficiently different from the now cleansed giver.

Daily special

Chicken schnitzel/Vichy carrots/potato wedges with cajun spice Dijon mustard sauce 900.

Mama mboga – pomodoro, mozzarella, spinach, carrot, baby marrow, sweet pepper 1200.

Green curry w/ prawns OR Green curry w/ vegetables 1500/1300.

Paradise cooler 650.

No diving

Written beside each pool. It is supposed to warn swimmers that the pool is not deep enough, that they might get hurt. It might also refer to in-depth political enquiry, like into the *Alliance Safari*, once owned by the former candidate for presidency, Matiba, who was tortured and is no longer alive. The bar keeper suddenly becomes silent. You are not allowed to touch the ground. The foundations of the resort, hotel, airport, sightseeing platform are obscured by plaster, decoration and poolscapes. Poolscapes are very bright, turquoise, always turquoise.

Nomad

Clients are now nomads. At the nomadic places they patronize, a sustainable lifestyle is possible. Eco projects, women's cooperatives, something to do with animals and the local community. A privileged social class responsible for extraction and ruination cannot bear to look at the outfall produced by their desire and consumption. Nomads are imagined as people who leave no trace but simply move through the savannah until they are out of sight.

Premium

Upgraded tourists travel in a way no way different from those on the average seats: they feel the same turbulences. What differs is the larger seat, the free champagne, the cotton napkin, the priority handling of their baggage. When such things don't work out as promised, the money invested in such class performance has been spent in vain. Waiting among the other passengers for luggage is downgrading. It is interesting that the verb *upgrade* is used frequently in global tourism, while opposite, *downgrade*, doesn't even seem to exist.

Return journey

By the time it is time to travel back home, we have just finished writing this book. All chapters are stored safely in their folders, and we share a delightful lunch to celebrate. We put our suitcases in the taxi and head towards Mombasa airport. Rain falls. The bypass that serves as the only access road, while the highway, intended to connect port, railway and airport, remains under construction, has turned into a muddy slope. Moving amidst the sliding trucks, we are still on time. Other tourists in other cars get stuck, as do the airline crew. The flight is delayed until the exhausted air hostesses also reach the airport. Going on for years now, the neglect and exploitation of entire regions results in ruination to the extent that finally even this breaks down: tourism on those beaches, the Kenyan beaches where part of our research took place, is on the decline, as everybody there said.

References

Ahmed, S. (2000) *Strange Encounters: Embodied Others in Post-Coloniality*. London: Routledge.
Aikhenvald, A.Y. (2004) *Evidentiality*. Oxford: Oxford University Press.
Akama, J.S. (2004) Neocolonialism, dependency and external control of Africa's tourism industry: A case study of wildlife safari tourism in Kenya. In C.M. Hall and H. Tucker (eds) *Tourism and Postcolonialism: Contested Discourses, Identities and Representations* (pp. 140–152). London and New York: Routledge.
Alim, S. (2016) Who's afraid of the transracial subject? Raciolinguistics and the political project of transracialization. In S. Alim, J.R. Rickford and A.F. Ball (eds) *Raciolinguistics: How Language Shapes Our Ideas About Race* (pp. 33–50). Oxford: Oxford University Press.
Allmendinger, P. and Haughton, G. (2009) Soft spaces, fuzzy boundaries, and metagovernance: The new spatial planning in the Thames Gateway. *Environment and Planning A* 41, 617–633.
Appadurai, A. (1988) Putting hierarchy in its place. *Cultural Anthropology* 3 (1), 36–49.
Appadurai, A. (1991) Global ethnoscapes: Notes and queries for a transnational anthropology. In R.G. Fox (ed.) *Recapturing Anthropology: Working in the Present* (pp. 191–210). Santa Fe, NM: School of American Research Press.
Ashcroft, B., Griffiths, G. and Tiffin, H. (1989) *The Empire Writes Back: Theory and Practice in Post-Colonial Literatures*. London: Routledge.
Atkinson, R.R. (1999) *The Origins of the Acholi of Uganda*. Kampala: Fountain.
Augé, M. (2001) *Les Formes de l'oubli*. Paris: Payot & Rivages.
Bachir Diagne, S. (2013) Truth and untruth: A conversation between Soxna and her friend Ngóór. In C. Jeffers (ed.) *Listening to Ourselves* (pp. 3–13). Albany, NY: SUNY Press.
Baker, M. and Kim, K. (eds) (2019) *The Routledge Handbook of Gastronomic Tourism*. London and New York: Routledge.
Beck, U. and Beck-Gernsheim, E. (2014) *Distant Love: Personal Life in the Global Age*. Cambridge: Polity Press.
Beer, B. (2000) Geruch und Differenz – Körpergeruch als Kennzeichen konstruierter 'rassischer' Grenzen. *Paideuma* 46, 207–230.
Belliveau, J. (2016) *Romance on the Road: Traveling Women Who Love Foreign Men*. Baltimore, MD: Beau Monde Press.
Berman, N. (2017) *Germans on the Kenyan Coast – Land, Charity, and Romance*. Bloomington, IN: Indiana University Press.
Blixen, T. (1937) *Out of Africa*. New York: Random House.
Bloch, A. and Donà, G. (eds) (2018) *Forced Migration: Current Issues and Debates*. Abington and New York: Routledge.
Block, D. (2014) *Social Class in Applied Linguistics*. Abingdon and New York: Routledge.
Blommaert, J. (2008a) Artefactual ideologies and the textual production of African languages. *Language & Communication* 28, 291–307.

Blommaert, J. (2008b) *Grassroots Literacy: Writing, Identity and Voice in Africa*. London: Routledge.
Blommaert, J. and Backus, A. (2011) Repertoires revisited: 'Knowing language' in superdiversity. *Working Papers in Urban Language and Literacies* 67, 1–26.
Bonfiglio, T.P. (2010) *Mother Tongues and Nations: The Invention of the Native Speaker*. Berlin: De Gruyter Mouton.
Boochani, B. (2018) *A Letter from Manus Island*. Adamstown: Borderstream.
Brantly, S. (2013) Karen Blixen's challenges to postcolonial criticism. *KULT* 11, 29–44.
Bruner, E.M. (1996) Tourism in the Balinese borderzone. In S. Lavie and T. Swedenburg (eds) *Displacement, Diaspora, and Geographies of Identity* (pp. 157–179). Durham, NC: Duke University Press.
Bruner, E.M. (2001) The Maasai and the Lion King: Authenticity, nationalism, and globalization in African tourism. *American Ethnologist* 28 (4), 881–908.
Bruner, E.M. and Kirshenblatt-Gimblett, B. (1994) Maasai on the Lawn: Tourist realism in East Africa. *Cultural Anthropology* 9 (4), 435–470.
Calacci, E. (2020) On acknowledgements. *American Historical Review* 125 (1), 126–131.
Cárdenas, I. and Guzmán, L.A. (2014) *Dólares de Arena [Sand Dollars]*. Santo Domingo: Aurora Dominicana.
Carman, J. (2002) *Archaeology and Heritage: An Introduction*. London: Continuum.
Carson, A. (2016) *Float*. London: Jonathan Cape.
Chiluwa, I. and Ajiboye, E. (2016) Discursive pragmatics of T-shirt inscriptions: Constructing the self, context and social aspirations. *Pragmatics and Society* 7 (3), 436–462.
Connell, R. (2007) *Southern Theory: The Global Dynamics of Knowledge in the Social Sciences*. Cambridge: Polity Press.
Corbin, A. (1988) *Le territoire du vide: L'Occident et le plaisir du rivage 1750–1840*. Paris: Flammarion.
Coupland, N. (2010) Welsh linguistic landscapes 'from above' and 'from below'. In A. Jaworski and C. Thurlow (eds) *Semiotic Landscapes: Language, Image, Space* (pp. 77–101). London: Continuum.
Cronopio, L. (2019) Backpacking performances: An empirical contribution. In A. Mietzner and A. Storch (eds) *Language and Tourism in Postcolonial Settings* (pp. 38–48). Bristol: Channel View Publications.
Cuno, J. (2012) *Museums Matter*. Chicago, IL: Chicago University Press.
Davis, A. (2015) *Simon's Town: Our Heritage*. Simon's Town: A. Davis.
de Guerville, A.B. (1906) *New Egypt*. London: William Heinemann.
de Lame, D. (2010) Grey Nairobi: Sketches of urban socialities. In D. Rodriguez-Torres and H. Charton-Bigot (eds) *Nairobi Today* (pp. 151–198). Dar es Salaam: Mkuki na Nyota.
Denbow, J. (2014) *Archaeology and Ethnography of Central Africa*. Cambridge: Cambridge University Press.
Deumert, A. (2014) *Sociolinguistics and Mobile Communication*. Edinburgh: Edinburgh University Press.
Dimmendaal, G.J. (2006) Kanuri. In K. Brown (ed.) *Encyclopedia of Language and Linguistics* 6 (pp. 154–155). Oxford: Elsevier.
Dixon, R.M.W. (2015) *Edible Gender, Mother-in-Law Style, and Other Grammatical Wonders*. Oxford: Oxford University Press.
Doyle, S. (2006) *Crisis & Decline in Bunyoro*. Oxford: James Currey.
Eckert, P. (2012) Three waves of variation study: The emergence of meaning in the study of variation. *Annual Review of Anthropology* 41, 87–100.
Etieyibo, E.E. (ed.) (2018) *Method, Substance, and the Future of African Philosophy*. London: Palgrave Macmillan.
Fabian, J. (1983) *Time and the Other*. New York: Columbia University Press.

Fabian, J. (2000) *Out of Our Minds: Reason and Madness in the Exploration of Central Africa*. Berkeley, CA: University of California Press.
Fanon, F. (2004) *The Wretched of the Earth*. New York: Grove.
Faraclas, N. (ed.) (2012) *Agency in the Emergence of Creole Languages*. Amsterdam: John Benjamins.
Ferreira, J. (2012) Colonial mimesis in lusophone Asia and Africa. See http://colonialmimesis.hypotheses.org (accessed 10 March 2015).
Gilroy, P. (2005) *Postcolonial Melancholia*. New York: Columbia University Press.
Gordillo, G.R. (2014) *Rubble*. Durham, NC: Duke University Press
Gramling, D. (2016) *The Invention of Monolingualism*. New York and London: Bloomsbury.
Greenblatt, S. (2010) Cultural mobility: An introduction. In S. Greenblatt, *Cultural Mobility: A Manifesto* (pp. 1–20). Cambridge: Cambridge University Press.
Gross, D.A. (2016) How the Swahili language took hold across Africa, and beyond. *The World in Words*, 22 November. See https://www.pri.org/node/155891 (accessed 7 February 2020).
Hachfeld-Tapukai, C. (2004) *Mit der Liebe einer Löwin: Wie ich die Frau eines Samburu-Kriegers wurde*. Cologne: Bastei Lübbe.
Hall, C.M. (1994) Ecotourism in Australia, New Zealand and the South Pacific: Appropriate tourism or a new form of ecological imperialism? In E.A. Cater and G. Lowman (eds) *Ecotourism: A Sustainable Option?* (pp. 137–158). Chichester and London: John Wiley.
Hall, C.M. and Tucker, H. (2004) Tourism and postcolonialism: An introduction. In C.M. Hall and H. Tucker (eds) *Tourism and Postcolonialism: Contested Discourses, Identities and Representations* (pp. 1–24). London and New York: Routledge.
Heller, M. (2011) *Paths to Post-Nationalism: A Critical Ethnography of Language and Identity*. Oxford: Oxford University Press.
Heller, M., Jaworski, A. and Thurlow, C. (2014a) Introduction: Sociolinguistics and tourism – mobilities, markets, multilingualism. *Journal of Sociolinguistics* 18 (4), 425–458.
Heller, M., Pujolar, J. and Duchêne, A. (2014b) Linguistic commodification in tourism. *Journal of Sociolinguistics* 18 (4), 539–566.
Henry, R. (2000) Dancing into being: The Tjapukai Aboriginal Cultural Park and the Laura Dance Festival. *Australian Journal of Anthropology* 11 (3), 322–332.
Honey, M. (1999) *Ecotourism and Sustainable Development: Who Owns Paradise?* Washington, DC: Island Press.
hooks, b. (1992) *Black Looks: Race and Representation*. Boston, MA: South End Press.
Huntgeburth, H. (2005) *Die weiße Massai* [*The White Masai*]. Frankfurt: Constantin Film.
Irvine, J.T. (2008) Subjected words: African linguistics and the colonial encounter. *Language & Communication* 28, 323–324.
Jaworski, A. and Pritchard, A. (eds) (2005) *Discourse, Communication and Tourism*. Clevedon: Channel View Publications.
Jaworski, A. and Thurlow, C. (2010) Language and the globalizing habitus of tourism: Towards a sociolinguistics of fleeting relationships. In N. Coupland (ed.) *The Handbook of Language and Globalization* (pp. 255–286). Oxford: Wiley-Blackwell.
Jay, S. (2018) The shifting sea: From soft space to lively space. *Journal of Environmental Policy & Planning* 20 (4), 450–467.
Jazeel, T. (2013) *Sacred Modernity*. Oxford: Oxford University Press.
Jennings, E.T. (2006) *Curing the Colonizers: Hydrotherapy, Climatology, and French Colonial Spas*. Durham, NC: Duke University Press.
John, E. and Rice, T. (1994) 'Hakuna Matata'. Los Angeles, CA: Walt Disney Records.
Kibicho, W. (2009) *Sex Tourism in Africa: Kenya's Booming Industry*. London and New York: Routledge.

Kiessling, R. and Mous, M. (2004) Urban youth languages in Africa. *Anthropological Linguistics* 46 (3), 303–341.
Kipling, R. (1990) *Something on Myself and Other Autobiographical Writings* (T. Pinney, ed.). Cambridge: Cambridge University Press.
Knapp, M. and Wiegand, F. (2014) Wild inside: Uncanny encounters in European traveller fantasies of Africa. In D. Picard and M.A. Di Giovine (eds) *Tourism and the Power of Otherness: Seductions of Difference* (pp. 158–175). Bristol: Channel View Publications.
Kopytoff, I. (ed.) (1987) *The African Frontier*. Bloomington, IN: Indiana University Press.
Kramer, F. (1987) *Der rote Fes: Über Besessenheit und Kunst in Afrika*. Frankfurt am Main: Athenäum.
Kreps, C. (2009) Indigenous curation, museums, and intangible cultural heritage. In L. Smith and N. Akagawa (eds) *Intangible Heritage* (pp. 193–208). London and New York: Routledge.
Larsen, K. (2008) *Where Humans and Spirits Meet: The Politics of Rituals and Identified Spirits in Zanzibar*. New York: Berghahn.
Lawson, S. and Jaworski, A. (2007) Shopping and chatting: Reports of tourist-host interaction in The Gambia. *Multilingua – Journal of Cross-Cultural and Interlanguage Communication* 26 (1), 67–93.
Leider, E.W. (2003) *Dark Lover*. New York: Farrar, Strauss & Giroux.
Leurs, K. (2017) Communication rights from the margins: Politicising young refugees' smartphone pocket archives. *International Communication Gazette* 79 (6–7), 674–698.
Lezra, E. (2014) *The Colonial Art of Demonizing Others*. London and New York: Routledge.
Lichtenberg, G.C. (1994 [1803]) Warum hat Deutschland noch kein grosses öffentliches Seebad? In *Schriften und Briefe III* (pp. 95–102). Frankfurt am Main: Zweitausendeins.
Lindtner, E. (2018) *Zwischen Nigeria und Europa: Schicksale von Migration und Remigration*. Vienna: Promedia.
Lipski, J.M. (2002) 'Partial' Spanish: Strategies of pidginization and simplification (from lingua franca to 'gringo lingo'). In C.R. Wiltshire and J. Camps (eds) *Romance Phonology and Variation: Selected Papers from the 30th Linguistic Symposium on Romance Languages, Gainesville, Florida, February 2000* (pp. 117–143). Amsterdam: John Benjamins.
Littlemore, J. (2015) *Metonymy: Hidden Shortcuts in Language, Thought and Communication*. Cambridge: Cambridge University Press.
Long, L. (ed.) (2004) *Culinary Tourism: Exploring the Other Through Food*. Lexington, KT: University of Kentucky Press.
Lorente, B.P. (2017) *Scripts of Servitude: Language, Labor Migration and Transnational Domestic Work*. Bristol: Multilingual Matters.
Lüpke, F. and Storch, A. (2013) *Repertoires and Choices in African Languages*. Berlin: De Gruyter Mouton.
Luttmann, I. (2007) Das Bild der Maasai aus der Sicht einer 'weißen Maasai': Kolonialer Erzählstoff in neuem Gewande? Filmrezension zu 'Die weiße Massai' (2005) von Hermine Huntgeburth. *Journal-Ethnologie.de*. See http://www.journal-ethnologie.de/Deutsch/_Medien/Medien_2007/Das_Bild_der_Maasai_aus_der_Sicht_einer_weissen_Maasai/index.phtml.
MacCannell, D. (1973) Staged authenticity: Arrangement of social space in tourist settings. *American Journal of Sociology* 79 (3), 589–603.
Mahenya, O. and Aslam, M.S.M. (2014) Rediscovering spice farms as tourism attractions in Zanzibar, a spice archipelago. In L. Jolliffe (ed.) *Spices and Tourism: Destinations, Attractions and Cuisines* (pp. 97–108). Bristol: Channel View Publications.
Maurer, R. (2015) *Fremdes im Blick, am Ort des Eigenen: Eine Rezeptionsanalyse von 'Die weiße Massai'*. Freiburg: Centaurus.
Mbembe, A. (2001) *On the Postcolony*. Berkeley, CA: University of California Press.

Médard, H. and Doyle, S. (eds) (2007) *Slavery in the Great Lakes Region of East Africa*. Oxford: James Currey.
Meiu, G.P. (2011) On difference, desire, and the aesthetics of the unexpected: 'The White Masai' in Kenyan tourism. In J. Skinner and D. Theodossopoulos (eds) *Great Expectations: Imagination and Anticipation in Tourism* (pp. 96–115). Oxford: Berghahn.
Meiu, G.P. (2017) *Ethno-Erotic Economies: Sexuality, Money, and Belonging in Kenya*. Chicago, IL: University of Chicago Press.
Melford, G. (1921) *The Sheik*. Hollywood, CA: Famous Players-Lasky.
Mietzner, A. (2019) Cameras as barriers of understanding: Reflections on a philanthropic journey to Kenya. In A. Mietzner and A. Storch (eds) *Language and Tourism in Postcolonial Settings* (pp. 66–80). Bristol: Channel View Publications.
Mietzner, A. (2021) Tomorrow means no: The transfer of Swahili linguistic practices into tourist settings. *Journal of Postcolonial Linguistics* 4, 13–29. https://iacpl.net/jopol/issues/journal-of-postcolonial-linguistics-42021/tomorrow-means-no/
Mietzner, A. and Storch, A. (eds) (2019) *Language and Tourism in Postcolonial Settings*. Bristol: Channel View Publications.
Mkono, M. (2011) The othering of food in touristic eatertainment: An ethnography. *Tourist Studies* 11 (3), 253–270.
Molz, J.G. (2007) Eating difference: The cosmopolitan mobilities of culinary tourism. *Space and Culture* 10 (1), 77–93.
Moore, P. (2002a) *No Shitting in the Toilet*. New York: Bantam Books.
Moore, P. (2002b) *Swahili for the Broken-Hearted*. New York: Bantam Books.
Murphy, R. (2010) *Eat Pray Love*. Los Angeles, CA: Columbia.
Nabhany, A.S. (1985) *Umbuji wa Mnazi*. Nairobi: East African Publishing House.
Namyalo, S. and Nakayiza, J. (2014) Dilemmas in implementing language rights in multilingual Uganda. *Current Issues in Language Planning* 16 (4), 409–424.
Nassenstein, N. (2016) Mombasa's Swahili-based 'Coasti Slang' in a super-diverse space: Languages in contact on the beach. *African Study Monographs* 37 (3), 117–143.
Nassenstein, N. (2017) Une promenade linguistique with a Senegalese street vendor: Reflecting multilingual practice and language ideology in El Arenal, Mallorca. *The Mouth* 2, 79–95.
Nassenstein, N. (2019) The Hakuna Matata Swahili: Linguistic souvenirs from the Kenyan coast. In A. Mietzner and A. Storch (eds) *Language and Tourism in Postcolonial Settings* (pp. 130–156). Bristol: Channel View Publications.
Nassenstein, N. and Rüsch, M. (2019) Skinscape souvenirs and globalized bodies: Tattoo tourism and language in East Africa. In S.T. Kloß (ed.) *Tattoo Histories: Transcultural Perspectives on the Narratives, Practices, and Representations of Tattoos* (pp. 237–255). New York and London: Routledge.
Nassenstein, N. and Storch, A. (in press) Balamane: Variations on a ground. In I.H. Warnke and E. Erbe (eds) *Macht im Widerspruch*.
Nyamnjoh, F.B. (2017) Incompleteness: Frontier Africa and the currency of conviviality. *Journal of Asian and African Studies* 52 (3), 253–270.
Ochieng'-Odhiambo, F. (2013) What's in a name? Four levels of naming among the Luo people. In C. Jeffers (ed.) *Listening to Ourselves* (pp. 52–89). Albany, NY: State University of New York Press.
Otsuji, E. and Pennycook, A. (2010) Metrolingualism: Fixity, fluidity and language in flux. *International Journal of Multilingualism* 7 (3), 240–254.
Penhaligon, J. (2011) *Speak Swahili, Dammit!* Falmouth: Trevelyan.
Pennycook, A. (2001) *Critical Applied Linguistics*. Mahwah, NJ: Lawrence Erlbaum.
Perbix, J. (2020) Alien(s) in paradise: Power relations, networks and the search for authenticity along Kenya's North Coast. Unpublished MA thesis, University of Cologne.
Perić, N. (2013) Corinne Hofmanns 'Die weiße Massai' und die gleichnamige Verfilmung von Hermine Huntgeburth. Unpublished diploma thesis, Sveučilište J.J. Strossmayera u Osijeku, Filizofski fakultet.

Philip, M.N. (1991) *Looking for Livingstone: An Odyssey of Silence*. Toronto: Mercury Press.
Phipps, A. (2007) *Learning the Arts of Survival: Languaging, Tourism, Life*. Clevedon: Channel View Publications.
Piller, I. (2016) *Linguistic Diversity and Social Justice*. Oxford: Oxford University Press.
Pollack, S. (1985) *Out of Africa*. Los Angeles: United International Pictures.
Rasmussen, S. (2016) *The Greatest Safari*. Pine Town: 30 Degrees South.
Reid, R. (2002) *Political Power in Pre-Colonial Uganda*. Oxford: James Currey.
Reid, R. (2007) *War in Pre-Colonial Eastern Africa*. Oxford: James Currey.
Richter, V. (2015) 'Where things meet in the world between sea and land': Human-whale encounters in littoral space. In U. Kluwick and V. Richter (eds) *The Beach in Anglophone Literatures and Cultures* (pp. 155–173). Farnham: Ashgate.
Roque, R. (2020) The name of the wild man: Colonial Arbiru in East Timor. In N. Nassenstein and A. Storch (eds) *Swearing and Cursing* (pp. 209–236). Berlin: De Gruyter Mouton.
Ruf, F.J. (2007) *Bewildered Travel*. Charlottesville, VA: University of Virginia Press.
Said, E. (1978) *Orientalism*. London: Routledge & Kegan Paul.
Salazar, N.B. (2004) *Developmental Tourists vs. Development Tourism*. New Delhi: Kanishka.
Salonen, S. (2014) *Patong Girl*. Berlin: Hanfgarn & Ufer Film.
Sarmento, J. (2016) *Fortifications, Post-colonialism and Power: Ruins and Imperial Legacies*. London and New York: Routledge.
Schneider, E. (2016) Grassroots Englishes in tourism interactions. *English Today* 32 (3), 2–10.
Seidl, U. (2013) *Paradies: Liebe [Paradise Love]*. Vienna: Ulrich Seidl Film Produktion.
Short, Y. (2004) *A Kitchen Safari: Stories & Recipes from the African Wilderness*. Cape Town: Struik.
Sindiga, I. (1999) *Tourism and African Development: Change and Challenge of Tourism in Kenya*. Farnham: Ashgate.
Skinner, J. and Theodossopoulos, D. (2011) *Great Expectations: Imagination and Anticipation in Tourism*. Oxford: Berghahn.
Smith, L. (2006) *Uses of Heritage*. New York: Routledge.
Smith, L. and Akagawa, N. (eds) (2009) *Intangible Heritage*. London and New York: Routledge.
Smith, L. and Akagawa, N. (eds) (2019) *Safeguarding Intangible Heritage*. London and New York: Routledge.
Smith, M.A. (2019) *Senegal Abroad*. Madison, WI: University of Wisconsin Press.
Stoler, A.L. (2009) *Along the Archival Grain*. Oxford and Princeton, NJ: Princeton University Press.
Stoler, A.L. (2013) Introduction. 'The rot remains': From ruins to ruination. In A.L. Stoler (ed.) *Imperial Debris* (pp. 1–37). Durham, NC: Duke University Press.
Storch, A. (2005) *The Noun Morphology of Western Nilotic*. Cologne: Köppe.
Storch, A. (2006) How long do linguistic areas last? Western Nilotic grammars in contact. In A.Y. Aikhenvald and R.M.W. Dixon (eds) *Grammars in Contact* (pp. 94–113). Oxford: Oxford University Press.
Storch, A. (2011) *Secret Manipulations*. New York and Oxford: Oxford University Press.
Storch, A. (2016) Communicative repertoires in African languages. *Oxford Research Encyclopedia of Linguistics*. See https://oxfordre.com/linguistics/view/10.1093/acrefore/9780199384655.001.0001/acrefore-9780199384655-e-155?mediaType=Article.
Storch, A. (2021) Shalimar. Language in the hollow body. *Journal of Postcolonial Linguistics* 4, 100–114. https://iacpl.net/jopol/issues/journal-of-postcolonial-linguistics-42021/language-in-the-hollow-body/

Storch, A. (in print) Waiting. On language and hospitality. In A.Y. Aikhenvald, N. Jarkey and R.M.W. Dixon (eds) *Integration of Language and Culture*. Oxford: Oxford University Press.
Storch, A. and Warnke, I.H. (2020) *Sansibarzone: Eine Austreibung aus der neokolonialen Sprachlosigkeit*. Bielefeld: Transcript.
Thiong'o, N.W. (1993) Her cook, her dog: Karen Blixen's Africa. In N.W. Thiong'o (ed.) *Moving the Center: The Struggle for Cultural Freedoms* (pp. 132–135). Oxford: James Currey.
Thurlow, C. and Jaworski, A. (2010a) *Tourism Discourse: The Language of Global Mobility*. Basingstoke: Palgrave Macmillan.
Thurlow, C. and Jaworski, A. (2010b) Silence is golden: The 'anti-communicational' linguascaping of super-elite mobility. In A. Jaworski and C. Thurlow (eds) *Semiotic Landscapes* (pp. 187–218). London and New York: Continuum.
Thurlow, C. and Jaworski, A. (2011) Tourism discourse: Language and banal globalization. *Applied Linguistics Review* 2, 285–312.
Tuhiwai Smith, L. (2012) *Decolonizing Methodologies*. London: Zed.
Urry, J. (1990) *The Tourist Gaze*. London: Sage.
Wacquant, L. (2017) Urban desolation and social denigration in the hyperghetto. In M. Mostafavi (ed.) *Ethics of the Urban: The City and the Spaces of the Political* (pp. 195–208). Zurich: Lars Müller.
Wainaina, B. (2005) How to write about Africa. *Granta* 92, 2 May. See https://granta.com/how-to-write-about-africa/.
White, H. (1973) *Metahistory: The Historical Imagination in Nineteenth-century Europe*. Baltimore, MD: Johns Hopkins University Press.
Wiszowaty, R. (2009) *My Maasai Life: From Suburbia to Savannah*. Vancouver: Greystone Books.
Yahya-Othman, S. (1995) Aren't you going to greet me? Impoliteness in Swahili greetings. *Text – Interdisciplinary Journal for the Study of Discourse* 15 (2), 209–228.
Yong-Hwa, K. (2006) *200 Pounds Beauty*. Seoul: REALies Pictures.

Index

African Frontier 87, 91
Arabic 6, 43, 106, 142, 153
authenticity 4, 8f., 108, 117f., 125f., 128, 130, 141, 144, 149, 151f., 155

Badaseraye, Festus 75
balconing 75
beach v, vi, viii, 2, 8, 9, 11f., 15ff., 30, 38, 42f., 67f., 72, 74, 78, 82, 84, 89, 91, 95f., 99ff., 105, 108, 110, 122, 126, 131ff., 155f., 160, 162f.
Beach Boy viif., 22, 95f., 98ff., 134
Binga 145f.
bingo 24f., 161
Blixen, Karen v, 12, 46f., 50ff.
Block, David 111f.
Boomu 157f.

Cairns 143, 156
Cairo viii, 105f., 119, 161
cannibal 22, 28, 35, 102
Cape Town 36, 83, 119, 142
Carson, Anne 56, 58, 60
charity tourism 13
Chopi 84ff.
Christiania vii, 37f., 152
click sound 56
coconut v, 13, 17, 21, 23ff., 68, 100, 124, 134, 150, 152f.
commodification 3, 35, 111f., 153
Connell, Raewyn 11, 50
consumerism 12, 15, 100, 122, 161
Corbin, Alain 18f.
Critical Heritage Studies 140
culinary tourism 12, 149, 157f.

dhow viii, 6, 21, 99f.
Diagne, Souleymane Bachir 76f.
Diani 8, 25f., 38, 42, 65, 68, 72, 100ff., 112, 115, 132

El Arenal 76, 108
Engelhardt, August 21, 28
English 2, 5f, 23, 32, 37, 43, 53, 73f., 82, 88, 95, 100, 107, 118ff., 129ff., 144, 151, 153
ethnophaulism 101

Fanon, Frantz 35, 122f.
film -> movie vi, 23f., 33, 37, 40, 42, 47ff., 51ff., 75f., 120, 122ff.
food 8, 12, 17, 24, 26, 83, 86, 112ff., 125, 132, 143, 149ff., 156ff., 161
Fort Jesus 5

German 2, 9, 22f., 33, 37, 50, 52, 74, 82, 86, 100ff., 107, 109, 118, 126, 130ff., 134f., 143, 152, 155
Greenblatt, Stephen 73, 77

Hakuna Matata v, 11, 23f., 26, 30ff., 59, 107f., 112ff., 134, 152, 155
Heller, Monica 3, 111f., 153
Helmut 75, 109
heritage ii, vi, 5, 48f., 72, 89, 138ff., 149, 151, 153, 155
Hone 146
hooks, bell 124f., 127
hospitality xi, 8, 18f., 24, 37, 49, 76f., 98, 153
hostility vi, 3, 102, 104, 110

imperial debris 48, 72, 94, 117
Indigenous knowledge 141
infinity pool 23, 96ff.
inscription 5, 8, 23, 36, 41f., 104, 152, 155
Irvine, Judith 116f.
Islam 76, 142, 146

Jambiani 10, 27
Jaworski, Adam 3, 18, 21, 95, 100
Jos 146

Karen (house) v, 46ff., 53
Katana, John 41, 44
Kenya 4f., 7f., 10ff., 22, 26, 31f., 33, 39, 43f., 48ff., 52f., 60f., 68, 88, 95f., 100, 112f., 115f., 119, 126, 128, 133f., (Kenyan 21, 23f., 26, 31, 33, 38, 40ff., 52, 60, 73, 82, 95, 119, 126f., 131, 134, 163 / Kenyans 40, 95, 100f., 132)
Kipling, Rudyard 83, 89f.
Kiryandongo 84ff., 90
Kopytoff, Igor 84, 87

language course vi, 31, 110, 112f., 119
Lichtenberg, Georg Christoph 20f.
linguistic landscape vii, 6, 111, 130
Lion King 32, 37, 41

Maasai/Masai vi, xi, 7f., 11, 82, 97, 108, 113, 115, 132ff., 148f., 155
Mallorca viii, 74, 105, 108ff.
Mami Wata 26
massage 16ff., 23, 25, 89
Meinhof, Carl 30
mimesis 76, 83, 90
mirative construction 58f.
Mombasa v, vii, 5, 12, 16, 25, 31, 39, 60f., 132, 135, 163
movie vi, 12, 34, 40f., 46ff., 50, 53, 121ff., 131f., 135
multilingualism 3, 84, 91
museum 5, 12, 42, 46, 48f., 51ff., 89, 138ff., 156

Nairobi 25, 46, 51, 97, 119, 126, 133
neocolonialism 130
ngalawa 21f.
Nigeria 74ff., 104f., 109, 146
noun class ix, 17, 24

Orientalism 84, 95, 108
Otium 19f., 24
Out of Africa 46, 50, 53

package tourism 10, 36
Paradies: Liebe (*Paradise: Love*) vi, 125ff., 133f., 134
Paradise 23, 71, 75, 95, 100, 127, 134, 162
Patong Girl vi, 125f., 128, 130
Persian 146, 154
Phipps, Alison ii, xi, 3, 18
Portuguese 5f., 43, 154

Postcolonial Theory 2, 4, 38f., 42, 51f., 94, 98, 122, 138f., 141, 149, 155f.
postcolony 149, 155, 160
pronouns 56, 58, 60, 117

repertoire 9, 22, 32f., 84ff., 100, 106, 143, 145, 153
resort 8, 12, 16, 18ff., 22ff., 59, 65, 68, 72ff., 82, 84, 95, 96f., 122, 131ff., 155, 161ff.
Richter, Virginia 20
rubble 17, 21, 48, 67f., 73
ruin vi, vii, 2, 6, 10ff., 19, 21, 67f., 71ff., 80, 130, 146
ruination 8, 10ff., 64, 72, 74, 130f., 145f., 163
ruinous xi, 4, 6f., 10, 20, 51, 117, 129f., 134

safari 10, 12f., 43, 49f., 52, 82ff., 89, 91, 97, 100, 112f., 116, 119, 149ff., 155, 162f.
Sand Dollars vi, 125f., 128, 130, 135
Seaford Town 143
Seidl, Ulrich 126f.
sex tourism vi, 12, 121f., 124f., 127f., 130, 133
silence xi, 7, 18, 60, 95, 102, 105, 129, 140
Simon's Town viii, 141, 142f.
smell terminology 17, 23, 83, 86f., 134, 150
souvenir 3, 12, 22f., 26, 32f., 36, 46, 74, 82, 100, 104f., 110, 112, 116f., 119f., 125, 152, 155
spa 15f., 18ff., 23ff.
speaker vi, 3, 10, 31, 58f., 84ff., 102, 105, 111, 120, 138, 144f., 160
spice tour 150ff.
spirit possession 26
subversion, subversive 8, 47, 51, 102 / 3, 27, 42, 52
Stoler, Ann Laura 6, 72, 145
Swahili vii, ix, 5f., 10, 17, 22ff., 31f., 35, 36, 38f., 40f., 43, 50, 82, 86ff., 108, 112ff., 134, 151ff.

t-shirt vi, viii, 3, 11f., 23, 35ff., 41, 74, 104ff., 117, 119, 162
taste 7, 83, 86f., 119 (tasteless), 133 (aftertaste), 149ff.

Them Mushrooms 31f., 40ff., 44
Thurlow, Crispin 3, 18, 21, 35, 100, 144
Tiwi vii, 60, 62, 65
Tjapukai 143f., 156, 158
Tuhiwai Smith, Linda 139f.
tuktuk 74

Uganda 10, 36, 84, 88f., 157
Ukunda 60ff., 132
UNESCO xi, 5, 72, 141

Valentino, Rudolph 123, 135
village tour 8

Wainaina, Binyavanga 4, 33
Walt Disney 32, 41f., 44
Weisse Massai (*White Masai*) vi, 125ff., 131
Westermann, Diedrich 30

Zanzibar viii, 6, 25, 27, 105, 107f., 112, 114, 116f., 120, 150ff., 158
Zar 25f.

For Product Safety Concerns and Information please contact our EU Authorised Representative:

Easy Access System Europe

Mustamäe tee 50

10621 Tallinn

Estonia

gpsr.requests@easproject.com